Putnam P Bishop

The Heart of Man

an attempt in mental anatomy

Putnam P Bishop

The Heart of Man
an attempt in mental anatomy

ISBN/EAN: 9783744736633

Printed in Europe, USA, Canada, Australia, Japan

Cover: Foto ©berggeist007 / pixelio.de

More available books at **www.hansebooks.com**

THE HEART OF MAN:

AN ATTEMPT IN MENTAL ANATOMY.

By PUTNAM P. BISHOP.

CHICAGO:

SHEPARD & JOHNSTON, PRINTERS,

140–146 MONROE STREET.

1883.

CONTENTS.

THE HEART OF MAN.

CHAPTER I.

I TAKE the word "heart" in the highest of its popular acceptations, and employ it to denote that part of the mental organism in which emotions are experienced, desires and affections exert their force, and volitions take place.

Already in my first sentence, there is an expression which would excite the contempt of many philosophers if they should happen to see it. They warn us against thinking of the Mind as made up of parts. They tell us that it is "unextended"—that it inhabits no space; and they would have us carefully guard against viewing it in the light of an organism. I must decline to acknowledge the authority of such teachers. However it may be to others, to myself, most certainly, an unextended entity is altogether unthinkable. I am compelled to employ the concept of space whenever I think of anything as having a substantive existence. To me, whatever exists in no space is nowhere, and what is nowhere is nothing. I do not aver, however, that Mind occupies space to the exclusion of Matter. Whether impenetrability is, or is not, a property of the mental substance, is a point on which I have no knowledge., But ignorant as I must remain concerning the nature of this substance, beyond the point at which its self-manifestations cease, I am under a necessity to postulate a "Something" in which the powers of the Mind inhere, and which affords a field for their movements and interactions. As names of an entity, such expressions as "A thread of consciousness" and "A series of 'feelings,'" convey no meaning. To talk of "consciousness" apart from a conscious subject, or of "feelings" without a subject that feels, is to talk nonsense.

I submit with perfect cheerfulness to .the necessities of which I have spoken, since all experience teaches that man is constituted for the discovery of truth. Hence, that an effort to cast off a necessary conception is a stride in the direction of foolishness, can be proved by an induction as broad as the whole range of human history.

Accepting the laws of my nature without reserve, I see no middle ground between the position that Mind is an extended and organized entity, and the position that all mental phenomena are products of our corporeal organism. Materialism may justly assert the prior claim to attention, for there is reason in the law which requires us, in all cases of this kind, to satisfy ourselves of the inadequacy of the causes which are known to exist, before assuming the existence of an additional cause. I have endeavored to obey this law. I have tried honestly to take in all the light which such men as Spencer, Bain and Maudsley are able to dispense. I am delighted and instructed by their disclosures concerning the structure of the brain and of the whole nervous system. While they linger in the domain of Anatomy, or in that of Physiology proper, I feel that I am following guides who know their way. I am deeply interested by their striking illustrations of the well known influence of corporeal changes upon the mental state, and of mental changes upon the corporeal state. But, so soon as they begin to deal with the simplest phenomena purely mental, I find an idea of inadequate causation growing upon me. When I place all the forces, which they point out, by the side of a complicated train of thought, with manifold emotions flashing out here and there upon it and interlacing themselves with it, my conviction that the former cannot have caused the latter becomes so overmastering that I lose all patience with the assertion of such origination. I am constrained to suspect that such able thinkers as I have named have busied themselves so exclusively with matter that they have become incapacitated for observations in which the bodily senses can render no aid. Be this as it may, their writings make it manifest that they have scarcely pushed their investigations across the threshold of mental life, and that they are profoundly ignorant as to the wonderful things which take place in the inner recesses of the human soul.

Of the countless *undisputed* products of material organisms, every one can be proved to consist of matter. Every one of them can be weighed, or measured, or discovered, by the microscope; or its presence can be detected by some other physical test. This being true, is the generalization unwarranted when we say "All products of material organisms are themselves material"? If any scientist has gone over a wider range in accomplishing his induction, I should like to hear his story. Assuredly some centuries will elapse before Evolutionists will gather a greater number of particulars to prove that all physical organisms are evolved from protoplasm. Perhaps it will be said that mental phenomena are not "products," properly so called, but *effects* akin to the corporeal feelings produced by an internal excitation of our nerves of sensation. I deny the validity of the distinction. I understand Physiology to teach that transformation of matter is a concomitant of all physical organic activity. Obviously, then, the new forms of matter, resulting from such activity, are immediate material products, and the consequent feelings are physically evidential of the presence of those products. Besides, a discriminating consciousness discloses so vast a difference between corporeal feelings and mental phenomena as almost to compel a recognition of different origins. The very least that it seems possible to say honestly is, that the differences between these two classes of phenomena are sufficient to establish an immense probability of difference in originating forces.

The materialistic theory can be antagonized legitimately with all the arguments which tend to sustain the authority of Christ as a Teacher, and with all other arguments going to prove the possibility of mental existence apart from a corporeal organism. It is very probable that the bringing in of such considerations will be called unscientific. I am venturesome enough to say, however, that if Science is what its name imports, the laying of all departments of knowledge under contribution, for the sake of getting at the truth, seems to me to characterize the only truly scientific method. At all events I shall never board up the windows on any side while endeavoring to adjust my conceptions to the realities of the universe.

For myself, I feel no need of support from the classes of arguments to which I have just made allusion. I hold that the human Mind is a living Being, because I can picture it to myself as existing apart from the human body, and can see it, in that state of isolation, exhibiting all those "changes, simultaneous and successive," which constitute a life complete in itself. The following chapters are largely taken up with the pointing out of such vital movements, and may be seen, from one point of view, to contain a continuous assignment of reasons for the rejection of Materialism.

The method of investigation, which I have employed in working my way to the conclusions set forth in this discussion, I call "The Anatomical Method," because it is based on the conception of Mind as an organized being, and aims to disclose the structure of that being. At the risk of appearing egotistical, I think it best, at this point, to relate an experience. About eighteen years ago I got tired of the vagueness of my ideas concerning conscience, and supposed I could soon clear them up with the help of the few authors accessible to me. I found myself instructed, of course; but the result, on the whole, was exceedingly disappointing. I came across no theory that was really useful in dispelling the mists in which my moral nature seemed to be enveloped. I then put aside my books and attempted, on my own account, to analyze the moral sentiments. I soon became convinced that "conscience," as the word stands in our literature, is the name of an ever-varying assemblage of wholly diverse elements, intellectual, emotional, etc., and ought never to be employed where scientific exactness is sought. As to the sources of those elements I was still in the dark, and found it necessary to go further back for a starting point. After several futile efforts I finally propounded to myself the question, What are the primary, undecomposable powers of the human soul ? Here I felt that I was making some headway, but had not proceeded far before it flashed upon me that I was treating the Mind as an organism, and ascribing different functions to its respective parts. I remembered, of course, that Sir William Hamilton says, "Man is not an organism." Nevertheless I persevered, and the farther I proceeded the firmer seemed the ground I was treading. From that time to this I have found great delight in conceiving the human Mind as an organized being, inhabiting a corporeal organism for the present, and affected immensely by that immediate environment, yet capable of living a life altogether its own. I have been conscious of no wavering in the opinion that the human Mind has a structure as definitive as that of the human body, and equally open to examination by one who is sufficiently resolute and able to hold his faculties to their work with sufficient steadiness. Hence I anticipate very sanguinely that the Mental Anatomists of the future will make good for Psychology a permanent footing among exact sciences.

Most of the contents of the following chapters were written out during the period of which I have spoken; and, after all these years, I see little to change except in the matter of terminology and in the way of condensation. It is possible that I am far astray in all my conclusions; but I cannot divest myself of the conviction that the lack of progress and of popular interest in psychological studies is largely due to the treatment of the mind as an unextended and unorganized "principle," or whatever else it is called, of which no conception can be formed. I believe, too, that a mere intellectual craving for a substantial and conceivable ground of mental phenomena has caused thousands of intelligent men to acquiesce in Materialism.

I distinguish in the mental structure five distinct systems, in each of which I see a distinct elementary power inhering. They are as follows:

1. The Sensory System; a system of mental senses.
2. The Intellective System; a system of faculties.
3. The Emotive System; a system of susceptibilities.
4. The Acquisitive System; a system of desires.
5. The Reactive System; a system of affections.

The elementary powers inhering in these systems, respectively, I call: Mental Sense, Intellect, Emotivity, Acquisitivity and Reactivity.

THE SENSORY SYSTEM.

Mental sense is the mind's power to take impressions from matter. It is that by which mind, as a recipient, is placed in correspondence with the material universe. In our present stage of being, the Sensory System is reached through our nerves of sensation and four special corporeal organs. If matter has classes of properties which cannot make themselves known through these avenues, the mind must remain ignorant of such properties so long as its inclusion in a material organism shall last. It is plain that we are not in a position to say that the entire Sensory System can now be brought under examination.

THE INTELLECTIVE SYSTEM.

The primary office of intellect is the ascertainment of truth. This involves the retention or permanent possession of truth, because continued ascertainment depends on the utilization of truth already ascertained. As I profess to treat only of the three systems which constitute the heart, I might, without impropriety, pass at once from this topic. I prefer, however, to say frankly that I have never attempted a thorough exploration of the Intellective System. My belief is that mental anatomists will hereafter make in the intellective part of our nature some important discoveries to which the ablest psychologists have never been able to attain in conceiving the mind as unorganized. If I were prosecuting researches in this field I should hold myself rigidly to the conception of the Intellective System as a system of FACULTIES, having TRUTH for their central object, and I should seek an exhaustive answer to this question: What are the structural provisions for the ascertainment, permanent possession, application, embodiment and expression of truth? I am satisfied that very insufficient study has been bestowed on what may be called " the executive processes " of intellect—the processes through which expedients are chosen, plans are constructed, machinery is invented and truth is made ready for expression in art and language.

There are three faculties, the workings of which are observed by everyone. I name them: the Intuitive Faculty, the Associative Faculty and the Introspective Faculty.

The existence of the Intuitive Faculty, as well as the fact of intuition, is, of course, denied by those philosophers who maintain that all our knowledge is derived from experience.

My patience with such men is hardly commensurate with the patience which would be elicited by a persistent endeavor to convince me that I lack a right arm. Some of them are blessed with wonderful endowments, but they are victims of a voluntary mania, which keeps them struggling to convince themselves and the world of the worthlessness of all our knowledge; and their work is uniformly pernicious. The Intuitive Faculty manifests its presence and power at the first dawning of infantile intelligence; it accompanies every attentive exercise of the organs addressed by the external world, and it is operative when the dying man bestows his last look on the faces of loved ones at his bedside. The instant a sensory impression is received, the Intuitive Faculty asserts the reality of that from which the impression proceeds. I venture here to throw out the remark that this last sentence seems to me to comprise a full statement of what takes place in sense perception. The ultimate facts are a sensory impression, and an intuition of reality. The interpretation of sensory impressions, by ascription of properties to objects perceived, is the work of other faculties, and is sometimes erroneous, while the Intuitive Faculty is infallible.

I find myself in full sympathy with Sir William Hamilton when he indignantly protests against certain doctrines because they imply that "the root of our nature is a lie." And yet the absolute and invariable trueness of our necessary beliefs is abundantly substantiated in another way. It is enough to know that, while false beliefs are plunging thousands in distress every hour, no man, in all the ages, was ever misled to his own harm by a necessary belief; that the Intuitive Faculty has so commended itself to all men that even those who deny its existence are ready, at any moment, to risk their lives on its veracity, and that the constant assumption of such veracity is an indispensable condition of safety. Let any man consider what would happen to him if he should contradict in practice the affirmations of the Intuitive Faculty as to Reality and Causation.

CHAPTER II.

THE EMOTIVE SYSTEM.

EMOTIVITY is the Heart's power to experience pleasure and pain from operations of Intellect. For each elementary form of pleasure, caused by a simple intellective action, the Emotive System presents an original Susceptibility. Otherwise, that form of pleasure could not be experienced. As pain is the opposite of pleasure, so each original Susceptibility of pleasure is, at the same time, a Susceptibility of pain from the opposite of that intellective action which causes the specific form of pleasure. Hence, in making a survey of the Emotive System, we have only to ascertain

THE PRIMARY INTELLECTIVE CAUSES OF PLEASURE.

These may be divided into two classes, according as the quality, relation, or state, which Intellect employs as a medium in the production of pleasure, is within the Mind, or external to it. In the former case, I call the Intellective Cause a Consciousness ; in the latter, a Perception.

Before entering upon a demonstration of the original Susceptibilities, I think it proper to remark that I believe an exclusive reliance on the Introspective Faculty, in psychological investigations, to be an immense mistake. As mental development progresses, the intellective processes, the emotions, desires and affections become more and more complex, and it becomes increasingly difficult for the Introspective

Faculty to reveal an elementary mental action. For this reason, I have relied largely, as will be perceived, on the observation of manifestations of mental life in childhood. At that period, there is no secretiveness to thwart us; lively emotions and strong impulses display their well known signs, and the general simplicity of mental action enables us to dispense with laborious analyses. I now proceed to what I regard as a complete list of the primary Intellective causes of pleasure.

1. *Consciousness of Knowing.*

I first used the word "Knowledge," instead of "Knowing"; but I saw that the emotion for which I was aiming to account, might be confounded with the self-complacency which a man experiences when he says to himself, "I possess a great store of information," whereas it is totally different. On the other hand, I have been, at times, inclined to name this cause simply "Knowing," and to treat the pleasure produced by it as merely the emotive reflex of intellective apprehension. It is certain that unmistakable signs of this special form of pleasure are observable before any keen introspection becomes possible. Still, I believe that, in the earliest stages of mental life, a certain dim thought of selfhood accompanies the emotion under consideration, and very soon a clear consciousness is manifested by such exclamations as "I see! I see!"

That Consciousness of Knowing is a primary cause of pleasure, will not, I think, be questioned. As the first inquisitive gaze of the infant evinces a spontaneous reaching of Intellect after truth, so the smile, with which its apprehensions are greeted, can be accounted for only on the hypothesis of an original susceptibility to that cause. As Intellect expands and its various faculties awaken to activity, if the tendency to reach after truth is judiciously stimulated and guided, the ebullient joy in learning names, apprehending distinctions and discovering uses, testifies still more strongly to the benign power of Consciousness of Knowing. Who has not observed the glee with which the little one points out the letters of the alphabet that have been learned, or attempts to name the articles of furniture in the room? Every intelligent mother understands the beguiling of infantile grief by asking "What is this?" and taxing her ingenuity to bring about new apprehensions. In later years Consciousness of Knowing goes into combination with other Intellective causes of pleasure, and, perhaps, is not often found apart from them. But the susceptibility to it continues to be a wellspring of joy, which needs only to be kept open in order to yield a copious stream to the end of life.

There are two circumstances by which the form of the pleasure thus produced is greatly affected. One of them is the order of apprehensions. When this is such that ideas akin to each other are raised successively, the emotion is like the tranquil flowing of a stream, more or less rapid, and continually deepening. But when apprehensions follow each other in such order, that strongly contrasted ideas are brought into being almost simultaneously, we have that titillation of spirit which is called the Sense of the Ludicrous, and finds expression in laughter. I would not be understood to say that this emotional state is occasioned in no other way. Laughter is often excited through peculiar operations of other Intellective Causes. But there are numerous cases in

2

which the emotions under consideration can be ascribed only to the order of apprehensions. We here discover a law evidently designed to give variety to our joys, and to refit Intellect for effective labor. Renewed elasticity and vigor are acquired in the fantastic mental actions occasioned by the combination of incongruous ideas, and by the presence of odd, grotesque and comical conceptions.

In the second place, the pleasure, caused by Consciousness of Knowing, is affected by the character of that which is known. This is a momentous fact. The comparatively unimportant truths, apprehended in early childhood, yield only a childish delight. The apprehensions, occasioned by gossip, cause only a petty pleasure. But the consciousness of knowing a sublime truth, a far-reaching relation, or a comprehensive law, raises, dilates and fills the soul with a joy beyond expression.

2. *Consciousness of Power.*

The first idea of power arises from physical action; and the earliest consciousness of it is the reflex of physical exertion. Very soon it begins to be measured by the distance passed over, the weight raised, or the resistance otherwise overcome. The memory of every one, who has seen children at play, will furnish abundant evidence of the title of this Consciousness to be ranked among primary causes of pleasure. As Intellect grows the Consciousness of Power attaches itself to intellectual exertion. Then it becomes intimately, though not inseparably, associated with Consciousness of Knowing, and doubles the pleasure produced by that Cause. When the period of planning and contriving arrives, a new field is opened. When, for example, a child first places a chair and climbs upon it, to take something from a table, a look of triumph evinces the delight which springs from consciousness of power to devise and execute that plan. The consciousness of knowing how to perform the feat may afterward present itself ; but I am convinced that, in most cases, Power is the sole medium of the first joy.

Again, we find Consciousness of Power uniting itself with the active impulses of the mind. When the impulse is pleasant in itself, the pleasure is heightened by a sense of its strength; and when it is painful the pain is greatly modified by the same cause. Our capacity to maintain fixed purposes affords a conspicuous field for this form of pleasure. One has but to search one's own internal history in order to be convinced that power of determination is, in itself, and without reference to its object, an important source of enjoyment.

Under this head, also, we may say that the amount and quality of the pleasure produced are affected by several circumstances. In the first place, other things being equal, the value of the emotion is governed by the order of the faculties from the exertion of which the consciousness proceeds. The lowest form is that which results from the exercise of physical power. Next comes the pleasure which is added to the product of Consciousness of Knowing in the mere acquisition of information. Far richer than this is the fruit of power in reasoning, in moving step by step, from deduction to deduction, toward a triumphant conclusion. Still more intensely delightful is Consciousness of Power to devise and combine expedients for the accomplishment of a great end. And yet higher than this is the pleasure yielded by that power which moves men to great achievements through the instrumentality of truth.

Again, the pleasure due to this cause is measured by the resistance overcome. This is merely saying that the resultant enjoyment is proportionate to the power necessarily employed. The law applies to every exercise of power, whether in the overcoming of physical resistance in the solving of difficult problems, in the influencing of men, in the regulation of one's own habits, or in the ruling of one's own spirit.

Finally, the other conditions being fulfilled, the value of the pleasure depends on the number of faculties brought into co-operative activity. Let the physical energies, Perception, Memory, Imagination, Reason and the faculties of expression be all aroused, each every instant throwing in its force to accelerate the one sweeping movement, and Consciousness of Power will then display the fullness of its worth. The emotions experienced on such an occasion are affected, of course, by the nature of the object aimed at; but, so far as this is the case, the result is traceable to other intellective causes. It should be noted that pleasure thus originated is not confined to the hours in which power achieves its triumphs. It is renewed when Memory recalls those hours, and is often anticipated while the triumphant moments are still in the future.

3. *Consciousness of Ownership.*

The term "Ownership" is chosen in preference to "Property," because it is broader and less liable to be misapprehended. I mean by it that relation to an object which enables me to speak of it as *mine*. That the consciousness of that relation is pleasurable, needs no proof. That it manifests itself in early childhood is equally unquestionable. The only point open to dispute concerns the possibility of its being included in Consciousness of Power. And I think all doubt on that point will disappear when it is considered that the right of exclusive control does not enter into the essence of Ownership. For reasons which could easily be explained, the idea of such a right is often associated with Consciousness of Ownership; and, in such cases, much of the resultant pleasure can be traced to Consciousness of Power. But we frequently use the possessive pronouns without the faintest notion of right or ability to exercise control. We say "*My* father," "*My* mother," "*My* party," "*My* church," "*My* country," and the delightful consciousness which finds expression in that emphatic "*My*," is plainly irreducible. There is certainly nothing in it which bears a near resemblance to anything in Consciousness of Power. I shall hereafter have occasion to show that the office of this intellective cause of pleasure is immensely important.

4. *Perception of Sympathy.*

I shall be understood, of course, to mean that we experience pleasure in perceiving Sympathy with ourselves. It is hard to explain how this cause of pleasure has escaped the notice of so many philosophers. They have much to say concerning Sympathy itself ; and some of them unduly magnify its influence in originating mental phenomena. But concerning the pleasure that rises within us at the manifestation of Sympathy with ourselves, they say little, if anything. No list of original impulses with which I am acquainted includes Desire of Sympathy. Yet one of the first things evinced by the infant soul is susceptibility of pleasure in finding its joys and sorrows shared by others. A little hurt is instantly changed to a benefit by that cause. All joys are multiplied by it. Much of the earliest prattle consists in calling attention to this and that, in order that apprehensions and emotions may be shared. In later years the susceptibility is continually manifesting its power. Our enjoyment of a landscape, a picture, a poem, a pleasing event, or a successful achievement, is incomplete till we are assured of Sympathy. The conception of this pleasure gives society its chief attraction, and is a principal stimulus to conversation, as well as to the advocacy of cherished opinions. The skillful management of the susceptibility has made the fortunes of many poets and other artists. After vainly endeavoring to trace these effects to some other Intellective cause, I deem it unquestionable that, if anything in man is original and ultimate, Susceptibility of Pleasure in Perception of Sympathy has that character.

5. *Perception of Favorable Regard.*

My reason for using the expression "Favorable Regard," instead of "Esteem," is obvious. The latter term is far too narrow, while the former includes affection, admiration and all other sentiments complimentary to their object. There can be no doubt that affection is the form of regard in the perception of which pleasure is first exhibited. The child is delighted by a loving embrace before a favorable opinion can be comprehended. Were it not for this, and one or two other facts, I should be strongly inclined to deny the originality of a Susceptibility to this Intellective Cause, and to ascribe the pleasure derived from Favorable Regard to a Perception of Sympathy with the perceiver's self-complacency, in connection with the strength added to the latter feeling by commendation. It is evident that commendatory manifestations often owe much of their pleasureableness to these influences. Yet, as I have intimated, there are some phenomena for which I cannot account without classing Favorable Regard among Primary Media of Pleasure. The matter would seem to be set at rest by the fact, that some persons are willing to exchange self-complacency for praise. Such is their conception of the value of Favorable Regard that, for the sake of it, they are willing to forfeit their own esteem.

6. *Consciousness of Goodness.*

A certain degree of mental growth necessarily precedes the manifestation of the Susceptibility corresponding with this cause of pleasure. The idea of personal quality must come into being, and a Mind must become able to reflect on its own character. But, so soon as this stage of development is reached, the Susceptibility is apparent to every thoughtful observer. The earliest qualitative discrimination is the act of distinguishing between Goodness and Badness. In the first instance it is objective ; external things are classed as good, or bad. This classification results inevitably from the reaction of Emotivity on Intellect. A perception is either pleasurable or painful, and, according as it is the one or the other, a favorable, or an unfavorable, judgment is immediately and spontaneously passed upon that which is perceived. The object then continues to be regarded as good, or bad, unless the judgment is reversed through subsequent operations of Intellect on Emotivity and the inevitable reactions of the latter on the former. When the child begins to reflect on its own qualities, a similar classification is made in these, and is corrected in a similar manner. All its qualities, physical, intellectual and moral, are regarded with pleasure or pain, and classed, accordingly, as good or bad.

Now, as favorable judgment is the fruit of pleasure experienced in the observation of that upon which the judgment is passed, it is self-evident that Consciousness of Goodness must be pleasurable. It may seem, however, that to predicate of this consciousness the production of pleasure, is to put the effect for the cause. But I think it will be found that the qualities, which please us when observed within ourselves, have, in every instance, been classed already as good by an objective discrimination. They have already addressed the soul and yielded pleasure in the Emotive System; and Emotivity has reacted on Intellect in causing a favorable judgment. We find, therefore, something exceedingly admirable in our being so constituted that, in the order of development, external observation shall precede self-knowledge. It is thus provided that our own qualities shall first come under our notice in association with judgments already passed upon them. The discovery in ourselves of a quality, already classed as good, is neither more nor less than a Consciousness of Goodness; and, as this Consciousness precedes and produces the pleasure we experience in regarding ourselves favorably, it is correctly set down as the Intellective Cause of that form of pleasure.

I see no occasion for argument in support of the view, that we have an original Susceptibility of Pleasure in Consciousness of Goodness. The pleasure thus originated is the emotional element in self-complacency. To be sure, this term is often applied to emotions produced by other Intellective Causes. A man is said to be self-complacent in view of his position in society, of the extent of his possessions, or of the power with which circumstances have clothed him. In these cases his pleasure is due chiefly to Perception of

Favorable Regard, Consciousness of Ownership, or Consciousness of Power. Strictly speaking, he has complacency in his
external condition, and not in himself. But the case is
entirely different when we experience delight in contemplating our own qualities, and calling them good, while all
thoughts of them in their relations to our other susceptibilities are excluded. That we are capable of such experiences,
can be denied by no one accustomed to introspection ; and
that this form of enjoyment is manifested in utmost simplicity by young children, is equally plain to every observer. It
will appear, in the course of this discussion, that Susceptibility to Consciousness of Goodness is an unspeakably valuable
endowment, since it lies at the very foundation, or rather,
since it *is* the very foundation of our Moral Nature.

In carrying forward our survey of the Emotive System,
we now meet with two Intellective Causes of Pleasure differing, in an important particular, from those already considered. They are perceptions of Media merely as objects of
contemplation, — Media having no special relation to the
perceiver ; neither existing within him, nor operating toward
him. The distinction is obvious. Knowledge, Power, Ownership and Goodness have been described as entering into
the state of the conscious Mind ; Sympathy and Favorable
Regard as having the perceiver for their object. But the
causes, which I am about to mention, may produce pleasure
without awakening a thought of self.

7. *Perception of Goodness.*

I have shown how the idea of Goodness arises, and how
the qualities of external things come to be classed as good or
bad. When I say that we are, by virtue of mental organi-

zation, susceptible of pleasure in Perception of Goodness, I simply assert that we are in such correspondence with the material and the spiritual universe, that numberless things apart from ourselves, and making no appeal to the susceptibilities heretofore pointed out, are capable of affecting us agreeably: certain sights, sounds, odors and tastes, and certain acts, relations and traits of character give us pleasure without producing a thought of anything pertaining to ourselves or tending to affect our condition. Such thoughts may be excited simultaneously by other Intellective Causes; but the enjoyment is highest when Perception of Goodness is the sole action of Intellect.

It will be observed that I use the term "Goodness" in its most comprehensive sense. It may be predicated of every being, every form of matter, every event, every action, every course of thought and every other conceivable subject of which the perception, without the intervention of any one of the other Intellective Causes, can produce delight in an unperverted Emotive System. The forms of Goodness are numberless. It embraces beauty, fitness, utility, rectitude, strength, and every other desirable quality and combination of qualities. Through the growth of Intellect all the primary Media of Pleasure come to be regarded abstractedly as good, though every Intellective Cause continues to have an independent operation.

At first the idea of Goodness shares in the vagueness of all other ideas. It is no more than a general favorable impression. But the intellective vision gradually acquires acuteness, and various distinctions are made. From the beginning, much of a child's attention is occupied with human beings. Very early, too, it shows itself favorably impressed by some, and unfavorably by others. It clings to one stranger and shrinks from another. It looks upon some as good, and upon others as bad. Then it begins to distinguish personal qualities, and becomes able to assign reasons for its likings and aversions. At a later stage of development the various forms of Goodness are classified. Physical Goodness, in the forms of beauty, grandeur, movements suggestive of power, etc., is seen to differ from intellectual Goodness. At length moral Goodness—Goodness of disposition, of intention, of volition and habit—is discovered to be distinct from every other species. Through all these stages of growth, Susceptibility to Perception of Goodness keeps pace with the intellective power of discernment.

8. *Perception of Happiness.*

There is a point of view from which this cause of pleasure may seem to be included in the preceding one. It may be said that happiness is merely Goodness of condition; and, therefore, that Perception of Happiness is only preception of a form of Goodness. But this position will not bear a moment's scrutiny. Both Happiness and Goodness may be predicated of the internal state of a rational being; and, however inseparable they may appear, we know that they are not one and the same. Both the conception and the emotion which we have when we say that a being is good are widely different from those which lead us to say that a being is happy. This matter is put at rest when we remember how often we have occasion at the same time to rejoice over the eminent Goodness of a man, and to deplore his unhappiness.

No one will deny that we naturally take pleasure in perceiv-

ing the Happiness of other beings. This is assumed intuitively
by every one. We try to please an infant by smiles and other
signs of joy in ourselves. The Happiness of children depends
very largely upon the evidences of Happiness which they witness in those around them; and this continues to be true of
most persons to the end of life. Though there is no limit to
the perversions possible within the human soul, it is hard to
find a man who has wholly lost his capacity to be pleased by
the pleasure of others. Envy, while intensely active may
paralyze the Susceptibility. It may disappear for the moment
under the force of reflections awakened by the contrast between one's own condition and that of another. Misanthropy
must nearly annihilate it. But the most hardened persons
show occasionally that they derive enjoyment from the Happiness of their fellow-men, and are pained by their unhappiness. It is notorious that a man who can no longer be moved
to acts of compassion instinctively shuns every scene of suffering unless he is impelled toward it by desire of gain or
by some other strong passion. While it is easy to see that
Perception of Happiness is more liable than other Intellective Causes of Pleasure to be rendered inoperative by the antagonism of selfishness, the perverting forces which it is able
to withstand bear witness to its original strength.

Although the facts above stated will be admitted universally, some will maintain that the pleasure ascribed to this
Cause is due to self-regard. In the view of many, the illusions
of Sympathy are boundless. Such will say that when we are
pleased by exhibitions of the happiness of other beings, we
simply imagine ourselves in their conditions. Others will
hold that in such cases we persuade ourselves that like happi-

ness is in store for us, and derive our enjoyment from that anticipation. I admit that Imagination is usually active in connection with Perception of Happiness; but its ordinary effect is only to render the perception more vivid. That it often amounts to an illusion, I do not believe. In scrutinizing my own internal histoŕy I cannot discover a single instance in which such a thing took place. I know, too, that I have found delight a thousand times in witnessing forms of enjoyment of which I had not the faintest expectation. An old man whose vitality is nearly exhausted and who moves about only with the greatest difficulty, sits in the doorway and observes the sports of his grandchildren. He is delighted by the exuberance of their animal spirits and by the joy indicated in their agile movements. Yet he neither imagines himself a child, nor expects to engage in such sports. He simply forgets himself; and, did he not do so, he would groan because of the contrast between his own physical state and that of the children.

The truth of the matter seems to be this: Every form of the pleasure perceived produces a certain likeness of itself in the Emotive System of the perceiver. It is reflected there as in a mirror. And as every emotion is a form of pleasure or a form of pain or a combination of the two, this imaging of the form is the beginning and the end of emotive Sympathy. Nor is there anything very peculiar in this. It is only a result of the general law: Pleasure shares in the modifications of its Media. We have seen that the pleasure which proceeds from Knowing is affected by the character of that which is Known, and that the emotive product of Power depends on the form of the Power apprehended in Consciousness. A similar

dependence of the form of the emotion may be asserted in
connection with all the primary Media of Pleasure. In the
present case, as Pleasure stands over against Pleasure, the resemblance is, of course, unusually distinct..

For aught that I can see, our examination of the Emotive
System is now complete. We have discovered eight original
Susceptibilities corresponding, respectively, with the same
number of Intellective Causes. That the names of these
Susceptibilities may be fully descriptive, I set them down as.
follows :

1. Susceptibility to Consciousness of Knowing, and Consciousness of Not-Knowing.

2. Susceptibility to Consciousness of Power and Consciousness of weakness.

3. Susceptibility to Consciousness of Ownership and Consciousness of Destitution.

4. Susceptibility to Perception of Sympathy and Perception of Lack of Sympathy.

5. Susceptibility to Perception of Favorable Regard and
Perception of Unfavorable Regard.

6. Susceptibility to Consciousness of Goodness and Consciousness of Badness.

7. Susceptibility to Perception of Goodness and Perception of Badness.

8. Susceptibility to Perception of Happiness and Perception of Unhappiness.

I have studied no simple emotion which cannot be traced
to one of these Susceptibilities, and no compound emotion
whose elements may not be found to have originated in two
or more of them. On the other hand, if this list of original

Susceptibilities can be reduced, I confess my inability to hit upon the method of reduction.

Perhaps I can obviate a little perplexity by explaining here that I give the name "Medium of Pleasure" to that with which Intellect busies itself in operating on the Susceptibilities. Knowing, Power, Ownership, Sympathy, Favorable Regard, Goodness, Foreign Goodness and Foreign Happiness, I call "Primary Media of Pleasure."

CHAPTER III.

WE can picture to ourselves a being with only the three systems already considered, and having no elementary power save Mental Sense, Intellect and Emotivity. Though such a being might present spontaneously all the internal changes which necessarily enter into a just conception of life, it would not be an agreeable object of contemplation. It could have no voluntary activity, and no impulse to exertion. It would be an aimless thing, drifting in the universe as chance might carry it. Its gratifications would be few and childish. Intellect, receiving no stimulus, the Intellective Causes of Pleasure would be feeble and limited to narrow ranges. We may find aid in forming a conception of the necessary state of such a being by abstracting, from their purely animal life, the indolent and aimless mental existence with which some men are content. But such is not the normal state of the human soul. It presents appearances of which we can see no explanation in the three elementary powers above named; and it is necessary, therefore, to continue our exploration.

One of the most conspicuous manifestations of mental life is a spontaneous reaching after objects, with a purpose to grasp them and reduce them to possession. This is the underlying and common characteristic of all desires. The power, thus spontaneously manifesting itself, I call Acquisitivity ; and I hold it to be an elementary power without which man would not be man. In my view, the action of an Intellective Cause of Pleasure upon Emotivity is followed by a reaching out of Acquisitivity after the Medium of the pleasure then experienced. Intellect and Emotivity originate the emotion, and Acquisitivity then demands a repetition, or a greater measure of it. Again, the Associative Faculty couples the peculiar form of pleasure with its Medium, and this, being the more definitely conceived, becomes the most conspicuous object of Acquisitivity. Such, I am convinced, is the genesis of all desires. Nothing is proved against this view by the fact, that we often desire what we have never possessed. In every such case the gratification desired is the fruit of Intellective Causes, of which we have had abundant experience. Moreover, Imagination provides for an association of specific enjoyment with a remote Medium quite as intimate as that which results from the actual possession of that Medium. At the same time, the fancied pleasurableness of the desired object often exceeds its actual gratifying power, because the pleasing features stand out in high relief while every drawback is wholly concealed. The origination of desire under such circumstances may be illustrated as follows : A man, who has always been poor, ardently desires to be rich. In the first place, he has experienced delight from Consciousness of Power, Consciousness of Ownership, and Perception of Favorable Regard. In the second place, under the impulsion of Acquisitivity, he has asked, "What would secure to me the combination of these delights in permanence?" Then Imagination has responded by picturing him in the possession of riches. He sees none of the cares and annoyances incident to that condition, but numberless

3

forms of pleasure have become associated with it in his conceptions. Thus an intense longing for wealth has arisen within him.

This theory of the genesis of desires is susceptible of abundant confirmation. It is obvious, for instance, to every observer of infancy that pleasure in each primary Intellective Cause precedes the desire of its Medium. The pleasure from Consciousness of Knowing precedes Desire of Knowing. If the impulsion of desire is inferred from the inquisitive gaze of the child, we can answer at once that the inference is unnecessary, since the spontaneous reaching of Intellect after truth sufficiently accounts for the phenomenon. It is still plainer that the delightful Consciousness of Power goes before Desire of Power. The same order is observable in the appearance of every desire of a primary Medium of Pleasure. The evidences of enjoyment are visible before any movement of Acquisitivity toward that which occasions it is indicated.

Again, we know that, in later years, many desires are actually formed in the manner described. Some persons instinctively shrink from gay society until they are brought by chance, or by constraint, to taste its attractions, and then become extravagantly fond of it. No one craves the excitements connected with dissipation unless he has previously fallen under their power. The great fear of parents is, that their sons will, in some way, be led into vicious indulgences and thus have originated within them perverse desires from which they are now free. All intelligent guides of the young endeavor to awaken elevated aspirations by occasioning elevated forms of enjoyment, and stimulating Acquisitivity to demand their repetition in greater intensity. In many such ways the correctness of our theory is constantly assumed.

Aside from these considerations, it is hard to see how any object can be desired before experience and the Associative Faculty have clothed it with pleasurableness. Our Consciousness testifies that, whenever we crave an object, we believe it capable of affording us enjoyment. We know that the association of pleasurableness with that object actually exists; and it would be altogether unphilosophical to treat that association as ultimate when it is so easy to trace it to previous interactions of Intellect and Emotivity.

According to the view here set forth, none of those impulses, usually classed as implanted desires, are underived. But it is equally clear that several tendencies of that nature are inevitably originated at a very early age. At the first experience of pleasure the action in the Emotive System is communicated to Acquisitivity, and the result is a vague Desire of Pleasure. Through subsequent experiences particular forms of pleasure become associated with the several Primary Media, and we have specific desires corresponding, respectively, with our original Susceptibilities, as Desire of Knowing, Desire of Power, Desire of Ownership, etc. These I call Primary Desires, and I am inclined to regard them as so many branches of that general Desire of Pleasure which is operative in every movement of Acquisitivity. Such a view affords no support to those who deny the possibility of unselfishness. They have been refuted already in our examination of the Emotive System. In the simple fact that, by virtue of its organization, the Mind is susceptible of pleasure from Consciousness of Goodness, Perception of Goodness and

Perception of Happiness, there lies and indestructible foundation for unselfishness. Besides, every one ought to understand that it is an abuse of terms to characterize any legitimate reaching of the soul after pleasure as *selfish*. That epithet is applicable only to *perversions* of the impulses necessarily originated by interactions of our elementary powers.

Desire of Pleasure, as we have seen, inevitably branches into eight Primary Desires. As the simple emotions, springing from the original Susceptibilities, are capable of an endless variety of combinations, and as the compound emotions thus produced may become associated with specific objects, it is plain that an unlimited number of specific desires can be permanently established in the Acquisitive System. Nevertheless, we continue through life to be often conscious of Desire of Pleasure in its original form. We have a vague longing for enjoyment without a distinct conception of any object as its Medium—a certain unrest of soul in which we are moved to say, "I want something; I know not what." We sometimes call it a craving for excitement; but it is obviously Desire of Pleasure not determined to any Medium, near, or remote. And here we discover an unspeakably important office of Acquisitivity. It is designed to urge us on to ever-enlarging possession of the Primary Media of Pleasure; and, therefore, it allows us no permanent repose while we are shrinking from the pursuit of those objects.

There is another form which this general desire assumes when life is prolonged for a few years. It unites itself with all anticipations, and, in that union, it is co-extensive in its reach with the associated conception of future existence. At length the idea, as it has been called, of "what is best on the whole" is formed, and immediately Desire of Happiness, in the proper sense of the name, springs into being. Jouffroy and others are right, therefore, when they term this "a desire of secondary formation," though its evolution, in a normally constituted Mind, is as certain and inevitable as that of any Primary Desire. There is propriety, likewise, in classing Desire of Happiness as a "rational desire," since its existence presupposes that contemplation of the distant future which is peculiar to rational beings. Its surpassing importance in the mental economy will be set forth hereafter. It is sufficient, for the present, to say that we should never lose sight of the distinction between Desire of Happiness and Desire of Pleasure. The vast difference between the two, as well as their co-existence, is abundantly manifested by their perpetual conflicts with each other.

Of most of the Primary Desires but little need be said at this time. Their existence will be admitted without extended argument; and it would not accord with my plan to dwell here upon their legitimate uses, or upon the perversions to which they are liable. I must call attention, however, to the fact that these desires are not connected with conceptions of the Primary Media of Pleasure as personal characteristics. Whenever a Medium is thus conceived it is a form of Goodness; and the desire of it is a mode of Desire of Goodness. For example, Desire of Knowing must be distinguished from a desire of *knowingness*—from a desire to be a *knowing* person. It is the impulse which moves young children to those perpetual questionings which so often tire the patience of parents and nurses. Sometimes it is called "curiosity," and sometimes "inquisitiveness." When operating singly, it is a desire to know merely for the sake of knowing, and is associated with no imagination of ulterior benefit. In like manner we are to distinguish Desire of Power from a desire of *powerfulness* as a trait of personality. When we are thinking of ourselves, and are conscious of craving power of body, of intellect, or of will, as an end, we are animated by Desire of Goodness. But when our characters are absent from our view, and we simply wish to exercise control in some particular direction, or desire to gain a certain position, in order that we may dominate, being intent solely on the gratification associated in our conceptions with the exertion of force, we experience Desire of Power in its simplicity.

Desire of Ownership so readily coincides with several other Primary Desires that it is found alone less frequently than some of the rest. Yet there can be no doubt of its separate existence in every human Heart. Say to a little boy that he may have a toy to play with whenever he likes, but that he must not call it his own, and he will be far from satisfied. If he has ever grasped the idea of possession, he will tell you "I want it for mine.". Any one, who will take the trouble of listening to children, will be struck with the peculiar emphasis with which they utter the words "my" and "mine," and with the fact that these are among their earliest utterances. And how shall we account for the conduct of the miser without recognizing Desire of Ownership as a separate impulse? He voluntarily forfeits all Favorable Regard, and refuses to use his wealth as a mean to Consciousness of Power. He subjects himself to manifold privations, and denies himself all influence, in order that he may hoard money and think of it as his own. It is certain that Ownership is an *end* to him, and not a mean to anything beyond.

It would be superfluous to prove the universality of Desire of Sympathy and Desire of Favorable Regard. Observation and Consciousness will furnish all who care to investigate such matters with abundant evidence.

If the genesis of desires is such as I have described, it follows that Desire of Goodness is entitled to a place among the primary and universal desires of mankind. We have seen that the Heart has an original Susceptibility of Pleasure in Consciousness of Goodness. Were it destitute of this, it would be incapable of self-complacency, in the proper sense of that term. But we know that such a capability is universal and is evinced at a very early age. And pleasure having once become associated with personal excellence, whether of body, of intellect, or of heart, Acquisitivity inevitably reaches out after that Medium. With every new conception of a form of personal Goodness, a new object of desire comes before the Heart; and thus the sphere of Desire of Goodness is enlarged.

When we seek to prove inductively the existence of this desire we are embarrassed by the fact, that Goodness, in numberless cases, soon comes to be regarded as a mean to the gratification of other desires. This result is due primarily to the law which conditions genuine Happiness on excellence of personality. It becomes apparent that the fruitfulness of all the Primary Media of Pleasure depends very largely on the worthiness of the individual. And the ordinary modes of parental training have a powerful tendency to

strengthen this conviction. All punishments are accompanied
by charges of Badness, and all rewards by commendations for
Goodness. Then children are told that, if they are good,
they will be loved and esteemed, and they have pointed out
to them the ways in which various excellences will work the
gratification of their various Primary Desires. The result is,
that Goodness ceases, in a measure, to be viewed as an end,
and comes to be regarded habitually as a mean. Consequently, when we examine cases in which Goodness is desired,
we may find many, in which it seems possible to trace the
phenomenon to other primary impulses, before we fall upon
one in which it is necessary to recognize a separate Desire of
Goodness. This fact goes far to account for the overlooking
of this impulse by many of the able investigators who have
surveyed this section of the human Heart.

Still, we need only to be thorough in our researches, in
order to be convinced, *a posteriori*, that Desire of Goodness
has a right to the place which I have assigned it. Observe
the children to whose minds Goodness has been carefully held
up apart from its consequences. Listen to their words as they
are giving expression to their anticipations. They will tell
you that they are going to do so and so, and add, " Then I
shall be good; shan't I ? " Question them after the evening
prayer, concerning their intentions for the morrow, and they
will tell you that they are going to be good girls, or good
boys, all day. It is evident that they are resting in the prospect of Goodness, without a thought of anything they are
to gain by it. And the delightfulness of the anticipation
proves the presence of Desire of Goodness; for it is impossible to anticipate with pleasure the realization of that which
is not craved.

We see the presence of this desire indicated more plainly, perhaps, in association with conceptions of particular forms of Goodness. The old and the young alike wish to be beautiful, to be strong in body and in understanding, to be kind and truthful and courageous ; and, although it often appears that these qualities are desired for the sake of Favorable Regard, or of some other consequence, yet in many cases it is equally plain that they are desired for their own sake. It may be remarked, also, that all persons, of every age, have an aversion to qualities which they consider bad, and desire to be free from them. Such freedom is desired on its own account ; and, though it is but a negative form of Goodness, it is plainly an object of the desire under discussion.

It is proper to ask here why it is that we are ever at pains to justify our conduct and our characteristics to ourselves. We often do so without a thought of what we are to gain or to lose by them. At such times we are evidently striving for the pleasure of thinking well of ourselves. But, if we did not desire to be good, why should we care to *think* ourselves good ? One who does not crave riches, never seeks to persuade himself that he is becoming rich ; and one, who has no thirst for fame or notoriety, never fights against the conviction that he is obscure. So, if we did not desire Goodness, we should never engage in solitary self-justification. I think every one is capable of an experiment by which this question can be settled conclusively. We can imagine ourselves outside the pale of moral government, where, as to the gratification of all other desires, it would be a matter of complete indifference whether we were good or bad. Now, supposing ourselves in that condition, would the prospect of becoming

mean, stingy, false-hearted and cowardly be as attractive as
the prospect of possessing all noble and heroic characteristics ? No one can make this experiment in good faith and
then doubt that he desires Goodness for its own sake.

If these evidences could all be explained away, I should
still maintain the primariness and universality of Desire of
Goodness. I know, that by reason of an original Susceptibility,
we take pleasure in Favorable Self-regard. I know, too, that
Acquisitivity is a power of every human heart, spontaneously
moving it to demand a repetition of every form of enjoyment
which it has once experienced. Hence I should believe in the
necessary origination of the desire in question, were I unable
to discover a single instance in which I could assert positively
that I had found it acting by itself.

There are two highly important desires which differ from
the other Primary Desires, as Perception of Goodness and
Perception of Happiness differ from the other Intellective
Causes of Pleasure. They demand the operation of their
Primary Media neither within the conscious Mind nor toward
the perceiving Mind. They simply call for the existence of
those Media as objects of contemplation. But these Media
are none the less fitted on this account to elicit the reaching
forth of Aquisitivity. I have shown that the right of exclusive control does not belong to the essence of Ownership. It
is equally apparent that exclusive enjoyment is not inseparable from that appropriation to which Acquisitivity impels.
When I bring an object into such relation to myself, that it
can operate through my Intellect to produce delight in my
Heart, I *acquire* it; and such an acquisition is a triumph of
desire.

These two Primary Desires are evolved by the conjunction of Acquisitivity with Susceptibility to Perception of Goodness, and with Susceptibility to Perception of Happiness; and I call them, respectively, Desire of Foreign Goodness and Desire of Foreign Happiness. It seems unnecessary to occupy much space in adducing facts to prove that these desires, through constitutional necessity, make their appearance in every normally constituted human Heart. Every child that remembers the beauty of an unclouded sky, and the glistening of the sunshine on the tree-tops, desires those forms of Goodness to reappear. When old enough to reflect on correctness of conduct, every child wishes to observe the signs of personal Goodness in its playmates. The preference of personal Goodness to personal Badness in others is seldom entirely lost by anyone. Occasionally one may be found who has progressed so far in mental suicide that he would be glad to drag others down to his own level. He is pained by the comparisons to his own disadvantage which are continually forced upon him; and the rage, engendered by Consciousness of Badness, impels him to seek relief in that way. But no thoughtful observer of men will believe that many persons have reached such a stage of depravity. On the other hand, we find many slaves of vice who sincerely and earnestly deprecate every step of another in the direction of their own moral state. As to the other desire, we may say that it is evinced habitually by every child, although it is sometimes overmastered temporarily by such an antagonistic impulse as Resentment or inordinate Desire of Power. Very few, if any, persons ever become so foolishly selfish that, other things being equal, they would not prefer the happiness of their fellow-men to their unhappiness, while, by a vast majority of human beings, that preference is manifested conspicuously, and under such circumstances as to necessitate the conviction that Foreign Happiness is desired for its own sake.

In these two impulses, differing so widely from the other Primary Desires as to the Media of Pleasure concerned in their evolution, we obtain our first view of the Heart's structural provisions for human beneficence. Desire of Foreign Happiness is by far the most powerful of the forces which impel men to mitigate human distress, and to engage in enterprises which have for their object the elevation of mankind. The co-operative beneficence of the two impulses is apparent when we consider that the end, toward which Desire of Foreign Goodness struggles, is an indispensable mean of human welfare.

As the Original Susceptibilities turned out to be dual, each of them presenting a possibility of pain as well as of pleasure, I at first expected to discover a similar duality throughout the systems which constitute the Heart, and to find a set of repugnances, acting as separate impulses, over against the several Primary Desires. In that case it would have been necessary to adopt a compound name for the Acquisitive System. It is true that, while our thoughts are occupied with the opposite of any Primary Medium of Pleasure, we are conscious of an impulse to seek freedom from it. Yet, as that impulse serves only to re-enforce the desire of the Medium itself, it is not entitled to be classed as a primary spring of action.

CHAPTER IV.

THE REACTIVE SYSTEM.

AMONG the actions of every child there are three particularly noticeable movements: an eager grasping, an open-handed caressing and an angry striking. The grasping is obviously a sign of Desire. The caressing and striking remain to be explained. This brings us to the examination of what I call the Reactive System, and of the fifth Elementary Power, to which I give the name Reactivity. This is the power which impels spontaneously to the reciprocation of pleasure and pain. By virtue of it the Mind spontaneously reacts, for the production of pleasure, on the person or thing it regards as the cause of its pleasure, and for the production of pain on the person or thing that seems to have caused a painful emotion in itself. It is the common characteristic of all affections, and bears the same relation to these that Acquisitivity bears to the Desires. This fundamental and obvious difference between desires and affections makes it a matter of astonishment to me that so many able psychologists have treated the two classes of phenomena as having the same genesis.

The Primary Affections are Love and Resentment. Every spontaneous impulse to reciprocate pleasure is a form of the one; and every spontaneous reaction of the Heart against pain is a form of the other. They are usually classed as, respectively, Benevolent Affections, and Malevolent Affections.

There is one distinction in forms of Love which is familiar to every one. When the causative pleasure is awakened by the bestowment of a benefit, viewed as intentional, we call the affection Gratitude. When the pleasure to which Love responds originates in Perception of Goodness, we call it Complacency. I see no better way than to adopt this latter name, though, in doing so, we must confine it rigidly to the disposition to confer. pleasure in return for delight experienced in Perception of Goodness. But there is another form of Love, more frequently exhibited than either of these, which seems to be still in want of a specific name. I refer to the affection which centers upon objects because they have become pleasurable through intellective association. For example, I am conscious of a strong attachment to a person who has never entitled himself to my Gratitude and whose character is not fitted to awaken complacency. In seeking the origin of this affection I discover, in the first place, that he has had several interesting experiences in common with myself; that I take it for granted that, in remembering those experiences, he has the same emotions with which I remember them, and hence that, whenever I think of him, I find pleasure in Perception of Sympathy. In the second place, I recollect that he has evinced a liking for me and thus contributed to my enjoyment through Perception of Favorable Regard. I recall, also, several instances in which I have exerted myself for his welfare and thus made him an occasion of pleasure though self-approbation, or Consciousness of Goodness. I now see that all these delights, blended in one, and dissevered from their causes, have become inseparably associated with my conception of that person, and that, consequently, the thought of him, whether present or absent, gives exercise to Reactivity. The form of the affection, here described, I call Associate Love. This, like many other names, which I have been compelled to introduce, fails to satisfy me entirely; but I can think of no better one. The above explanation will, I trust, preclude any confusion which might otherwise result from the ambiguity of the expression.

It is not often, perhaps, that one of these forms of Love exists for a long time apart from the others. Gratitude gives pleasure to the grateful Heart through Consciousness of Goodness; and in this way, if in no other, becomes combined with Associate Love. It also predisposes one to discover or to imagine, in its object, characteristics which may properly elicit Complacency. This latter effect is produced, also, by Associate Love. In its turn, Complacency, being itself delightful, is followed by an additional association of delight with the conception of the object toward which it is directed. Moreover, both Complacency and Associate Love predispose Intellect to ascribe intentional beneficence to their object, and thus to evoke, through Emotivity, an accession of Gratitude. In a majority of cases, therefore, though one of the forms is usually far more conspicuous than either of the others, the affection is compounded of Gratitude, Complacency and Associate Love. While I see these wonderful interactions taking place, it would be hard to keep my pen from setting down an expression of admiration for the structural provisions whereby the dwelling of Love in the

human Heart is made sure, or from charging impudence
upon the man who asks me to believe that all these manifestations of a mental life complete in itself have their
primary and sole origin in the gray matter of the brain, or
in a "thread of Consciousness."

With the facts now before us it becomes easy to account
for any special affection. Friendship, for instance, may have
any one of the three forms of Love for its original and chief
component. Sometimes it is, at first, a mutual complacency.
In other cases it is originally, on the one side, Gratitude for
benefits received, and, on the other, Associate Love called
into being, through consciousness of Goodness and Perception
of Happiness, in the bestowment of those benefits. More
frequently, I think, it begins on both sides in Associate Love.
This form of Love, too, is the chief element of affection for
kindred, as distinguished from friendship. Here, however,
the most important part is filled by an Intellective Cause of
Pleasure which I have not yet mentioned in this connection.
I refer to Consciousness of Ownership, and it was with an
eye to its power in shaping the affections, that I spoke in a
former chapter of the great importance of this cause. The
ability to say of a person "He is *my* father," or "He is
my son," often yields an intenser delight, and, therefore,
causes a stronger movement of Reactivity than all other
forces combined. This truth was dimly perceived by Stewart,
when he said, "The parental affection takes its rise from a
knowledge of the relation in which the parties stand." To
the Consciousness of Ownership — of the ability to use the
possessive pronouns of the first person — it is due that parents
and children, brothers and sisters and more distant relatives

may have an ardent mutual affection, though they have had few experiences in common, and are indebted to each other for little pleasure through any of the other Intellective Causes. It is to this alone that we are to ascribe the difference between friendship and affection for kindred when the other causes of Complacency, Gratitude, and Associate Love have been equally operative in the two cases.

It will at once occur to the thoughtful reader that Patriotism is, in the main, the response of the Heart to the pleasure received, in the contemplation of one's country, through Consciousness of Ownership. To be sure, Complacency always enters into the affection. The Love originated by the former Cause, determines Intellect to dwell upon all that is good, and to ignore all that is bad, in the territory and institutions of one's country, and in the national characteristics of its inhabitants. The pleasure is intensified and Reactivity is stimulated by remembrance of all that was heroic in the achievements of one's countrymen in past generations. Intellect is moved, likewise, to occupy itself with the personal advantages for which one is indebted to one's country ; and thus Gratitude, also, is brought to the support of Associate Love. Still, a man loves his country primarily because it is *his* country. But for Consciousness of Ownership there would be no such distinct affection as patriotism in the world. Similar remarks are applicable to Love of Party. This affection, having once originated in Consciousness of Ownership, disposes one to keep in view what one believes to be good in the history, characteristics, measures and principles of one's party, and thus goes into combination with Complacency. The disposition to magnify all that can be made to seem good, and to

4

shut all that is evil from view, often becomes very strong
through the vehement exercise called forth by opposition;
and Associate Love receives additional fervor from the pleasure produced by Consciousness of Power in the anticipation
of party triumph.

It may seem, at the first glance, that there are some affections which cannot be accounted for on this theory. For
example, one may say, "Here is a man laboring for the welfare of a class of vicious persons. He evidently loves them.
Yet he has never had any relations to them which could give
rise to Associate Love; there is nothing in their characters
to evoke complacency; certainly they have never earned his
gratitude. How comes it about, then, that he loves them?"
In answer to such an inquirer I would call attention to the
distinction between Desire and Affection. There is peculiar
danger of confusion at this point, since the movement of
Reactivity, originated by pleasure, coincides precisely with
Desire of Foreign Happiness. These two distinct provisions
for human beneficence belong to the structure of every human
Heart. But is it not plain that, in the case above instanced,
the benefactor was moved at the outset by Desire of Foreign
Happiness and Desire of Foreign Goodness? It is probable
that now those impulses have been reinforced by some measure of affection. His labors have yielded him pleasure in
Consciousness of Goodness. It may be, too, that he has found
gratification in Perception of Favorable Regard and in Consciousness of Power. We can easily see, therefore, how
Associate Love may have been generated since those Desires
became operative. Moreover, in most cases of this kind, it
will be found that the attention of the philanthropist has

fixed itself on some good quality — some redeeming trait — in the characters of those for whom he is laboring, and, therefore, that complacency is not wholly wanting. It may be well here to emphasize the truth that simple benevolence, such as we feel toward mankind at large, *is not an affection*. It is neither more nor less than Desire of Foreign Happiness. There is one singularly complex phenomenon to the origination of which all the Elementary Powers contribute. It is Pity, and the analysis of it is not difficult now that we know what our Elementary Powers really are. Mental Sense receives an impression which Intellect interprets as a sign of suffering. This interpretation causes pain though Susceptibility to Perception of Unhappiness. The influence of this pain, extending to the Acquisitive System, stimulates Desire of Foreign Happiness to set the intellective faculties at devising means of relief. Then a pleasant emotion is awakened through Consciousness of Goodness; and, finally, this pleasure is associated with the person pitied, and Reactivity yields Associate Love. If the Pity is prolonged, the affection is strengthened in various ways which it is unnecessary to specify.

Love is often combined with a desire for the presence of its object, and is sometimes viewed as including it. But this craving is no part of the affection itself. Enjoyment having become associated with the presence of the object of Love, Desire of Pleasure accounts for all the rest. We long to be at home because it is associated in our conceptions with delight, and we desire to be delighted. For the same reason, the devout person is attached to his accustomed place of worship, and all persons desire to revisit the spots where they have experienced great pleasure.

It is true, indeed, that we do sometimes *love*, in the strictest
sense of the term, persons, places and things that we are unable to benefit. But this by no means disproves our theory.
The impulse to reciprocate the pleasure received from them
still remains. The force of Reactivity in such cases is often
sufficient to produce a momentary illusion in the most clear-
sighted intellect. We find ourselves thanking and blessing
the writers and doers of former ages in return for the delight
which their works and deeds have given us. We feel that we
must, in some way, gladden them with a knowledge of our
appreciation. Who has not found himself caressing a tree,
or stretching out his hands to his native hills, with the same
feeling that accompanies the reciprocation of pleasure ? Who
has not pronounced benedictions on the landscape that was
filling him with joy ? In fact, I believe that the force of Reactivity is most fully brought to light in some of those cases
in which the bestowment of benefit is impossible. What is
the meaning of adoration, if the Heart has no impulse to re-
act, for the bestowment of blessedness, toward One whose
blessedness cannot be increased ? There is no anomaly here,
nor is any blunder of creative power indicated by these
things; for, in this part of its functions, Reactivity sheds
back upon the Heart the benefits which it is unable to com-
municate.

I see no reason to dwell at length on Resentment. Its
uses and possible perversions, as well as those of Love, and of
all the Primary Desires, afford a most inviting field for investigation. But my present aim is simply to point out what I
find in the human Heart; and I desire to make this discussion
as brief as possible. The forms and manifestations of Re-

sentment are the opposites of those of Love. Opposed to Gratitude we have Revenge; opposed to Complacency we have Indignation, and opposed to Associate Love we have Associate Hatred. The Geneses of the forms of Resentment take place according to the same laws under which the forms of Love are originated ; and there are similar comminglings, resulting in peculiar exhibitions of ill-will. As pleasure operates on the one side, so pain operates on the other.

One statement has been repeated by so many writers that it ought to be corrected. I mean the assertion that Resentment is invariably painful. To be sure, it always originates in pain, and while the facts which occasioned it are kept in view, a degree of pain must continue to be felt. But these things do not prove that the affection is at all times, on the whole, a painful one. Among savages it is even a proverb that " Revenge is sweet; " and, where the passion is universally regarded as a virtue, we can readily believe that this is true. In Christian lands, of course, the case is different. Here every one understands that deliberate Revenge, operating singly, is evil, and consequently it must come in conflict with Desire of Goodness as well as with Desire of Foreign Happiness. Every one believes, also, that it is against his own welfare to cherish an evil affection, and, hence, a struggle arises between Revenge and Desire of Happiness. Envy and jealousy must always be painful, because they are obviously perversions of Resentment. But this is far from being true of just Indignation. I hold that when a disposition to favor the infliction of punishment has been aroused by a gross manifestation of badness, and is, therefore, in accord with Desire of Goodness, and when the Happiness of the offender

is duly subordinated by Intellect to the welfare of society and
Desire of Foreign Happiness is thus brought into co-opera-
tion, and when the disposition is not thwarted by the hinder-
ance of justice, the pleasure experienced far exceeds the pain
that enters into the state of mind. I remember a Sabbath
morning on which it was announced that one, believed to be
a great criminal and responsible for incalculable distress, had
been arrested ; and I can recall few occasions on which I have
witnessed equal manifestations of popular joy. Yet Resentment was the one impulse which predominated over all others
when that man was the object of thought. It seems to me
unquestionable that the legitimate working of Reactivity in
the direction of pain is designed to increase the sum of our
enjoyments.

CHAPTER V.

We have traced the geneses of Emotions, Desires and Affections, and have seen how numberless motives to exertion come into being. We have yet to observe the processes through which these springs of action become effective. We must see them at their work; and, in order to make my demonstration clear, I find it necessary once more to enlarge my nomenclature. Desires and Affections, both primary and specific, considered as operating forces, I call Impellents; and I shall have occasion to apply the name "Impulse" to the force resultant from the conjunction of any two, or more, of these impellents, though I shall continue sometimes to use it in its ordinary sense. The connection, I hope, will show with sufficient clearness what I mean by the word.

That which takes place just before an impulse operates on Intellect and sets it in motion, is one of the most conspicuous manifestations of mental life. Its well-known name is "A Volition;" and my definition of it is this : A volition is the rise of an impulse to operativeness. That impulse may be a a single impellent, primary or specific ; or it may be the resultant of several impellents co-operating, or conflicting. In either case I think it clear that its rise to operativeness does not presuppose the activity of an original faculty distinct from the Elementary Powers already examined.

In the present state of opinion on this subject I feel myself called upon to occupy some space in defending the position just stated. Most persons ascribe volitions to what they call "The Will," and treat this as a distinct and underived faculty of supreme importance. As President Hopkins says : " By some, by most indeed, this element of will is supposed to be the chief one in personality, and there are those who regard it as the only one." It may be expected that those who share this view will be indisposed to acquiesce in the conclusion that all volitions can be traced to the interactions of the Powers which I have described. Yet with all deference I am prepared to affirm that I neither feel the need nor am able to discover the existence of a separate faculty possessing the attributes usually ascribed to the Will. My view is that all the striking phenomena, connected with volition, originate in movements of the impellents preliminary to contact with the executive powers of Intellect. They result from the more or less obstructed operation of Acquisitivity and Reactivity toward Intellect. Nor do I see any reason for interposing a separate faculty between the two former and the last, which would not equally call for such an interposition between any two of the Elementary Powers, in order to account for the effect of the one upon the other. Take an illustration. Intellect assumes the form of Consciousness of Knowing, and thus excites pleasure in the Emotive System. Emotivity now acts upon the Acquisitive System, and produces Desire of Knowing. By the occurrence of a problem this primary desire becomes specific : I desire to know the true solution of the problem ; and instantly this specific desire acts upon Intellect and turns it into the mode of investigation. I am voluntarily concentrating my thoughts and endeavoring to unearth the truth buried in that problem. Now, how can it be made to appear that in this case the desire does not affect the intellective movement as immediately as Consciousness of Knowing acts on Emotivity?

It is true that we are often more vividly conscious of the contact of Acquisitivity or Reactivity with Intellect, than we are of the previous operations of the Elementary Powers on each other. This is not true in every case. In the one just instanced, and in a thousand others which might be adduced, the last conjunction attracts no more attention than is given to any one of those which precede it. And we can easily explain why it is ever otherwise. The final contact of the impellents with the executive powers of Intellect is often the result of a protracted struggle. The Mind is conscious of a number of discordant impulses, no one of which is enough stronger than the others to gain an immediate victory. A period of hesitation and wavering ensues. In the meantime, Intellect is engaged in comparing, imagining and making various other movements, all of which announce themselves in Consciousness. Finally, the judgment, "This is best," is pronounced, and simultaneously the resultant impulse becomes operative. In this way the rise of what we call voluntary action becomes a conspicuous object of attention; and the contact, which is its immediate origin, is strikingly distinguished from the preceding ones. It is probable that this vividness of consciousness, in connection with volitions, is one of the chief reasons why men are so strongly disposed to ascribe them to an independent faculty. The phenomena are very conspicuous, and one naturally desires to account for them. In such cases the shortest way to dispose of a troublesome question is to assume that an original faculty has been implanted for the special purpose of yielding the results under consideration.

Another circumstance, which draws attention to the last step in the origination of activity in the executive powers, and heightens the estimate of its relative importance, is the appearance of its decisiveness in regard to conduct. The volition, being immediately connected with the act, is regarded as its sole cause. This conclusion results from a hasty suspension of investigation. When we look farther back we discover other decisive moments at earlier stages of the process. In the case supposed above, if Consciousness of Knowing had produced but a slight effect in the Emotive System, the acquisitive impulse would have been too feeble to move Intellect to the solution of the problem. Again, though Desire of Knowing had been strong, had Acquisitivity, at the same time, been excited still more powerfully through some other cause, the latter impulse would have controlled the Intellective movement, and there would have been no attempt to solve the problem. We see, therefore, that the contacts of Intellect with Emotivity and of the latter with Acquisitivity were quite as decisive as to the volition itself as the latter was in regard to the subsequent act. This conclusion is amply justified by observation. We know that a person, in whom Desire of Knowing is strong, habitually sets himself about the solution of problems which have no attraction for one in whom that desire is weak. We know, too, that the one who has the desire in great strength experiences incomparably more pleasure than the other in Consciousness of Knowing; and this greater pleasure obviously accounts for the greater strength of the desire.

A difficulty may be suggested here in connection with the fact, that the volition is not always followed immediately by the corresponding action. I may will to do a thing at a future time. In this case there is evidently a volition, which takes the form of a purpose, though there is yet no contact with the executive powers. Still, so far as the mental state is concerned, the preparation for that contact is complete. The impulse, too, may properly be called "operative," because it has gained predominance over all opposing impulses, and only awaits the passing of obstructions in order to exert its power. For its preservation in that state there is need only of that general power of the Mind to retain its tendencies, which is one of the indispensable provisions for the continued existence of the Mind as an organized entity. Nor is there anything singular in the pause which precedes the contact of the operative impulse with Intellect. Similar pauses often take place before other operations of Elementary Powers on each other. Take three cases. Ten years ago, Desire of Knowing became specific, and rose to the state of an operative impulse to investigate a certain subject. It was necessary to defer the investigation; but today I apprehend a favorable opportunity, and the operative impulse sets my faculties at work. Ten years ago, I saw a picture which gave me intense delight through Susceptibility to Perception of Goodness. I had not the faintest desire to possess that picture, because I supposed its acquisition impossible. Today, I apprehend the truth that I can make the picture mine, and am immediately conscious of desiring it most intensely. Ten years ago, I saw a person in great distress, on account of the act of another. I felt no resentment, however, because I considered the act accidental. Today, I apprehend the truth that the injury was inflicted wilfully ; and I am possessed at once with vehement indignation. In each of these cases an intellective apprehension breaks up a pause of ten years' duration ; and I see no more need of an intermediate faculty in the first case than in the others. To all such transitions of vital force from one System to another, there are indispensable conditions, which, if not present, must be awaited.

"But," it may be asked, "does not Consciousness, after all, testify to the existence of an independent Will ? Is there not something in the consciousness of a purpose which cannot be produced by the process described ? When I say, I *will* do this, have I not a feeling which must have come from some other source ? " In answer to such questions, I would say, in the first place, that Consciousness alone delivers no testimony concerning the sources from which our feelings spring. It busies itself exclusively with the processes of the moment, and discloses absolutely nothing as to their origins. It is equally plain that Consciousness presents no analysis of a feeling compounded of several elements. It merely provides material for the exercise of analytic power; and the Consciousness of a feeling, which it is exceedingly difficult to resolve into its component elements, is a matter of hourly experience. Several emotions, entirely distinct from each other in their origins, are blended in one ; and the compound is widely different from any one of the components. The peculiar consciousness, resulting from such a combination, can be known only by its actual presence in the Mind. Without previous experience in that very matter, and a perfect analysis of that very mental state, and a tracing of each component emotion to its source, it is impossible to say that the blending of certain specified mental processes would, or would not, produce a given effect in Consciousness. Hence, the feeling which we have when we say "I *will* do this," neither affords any evidence concerning its own origin, nor proves the impossibility of its origination in the manner I have traced.

Let us glance again at what is involved in the rise of an impulse to operativeness. It is unnecessary to dwell on the case of a volition preceded by no conflict, and immediately going into effect. Of such a process, as I have said, the consciousness is no more vivid than that of any other operation of one Elementary Power on another. But what takes place when there is a struggle? There is a comparison of the objects toward which the several impellents are pressing. Imagination places us successively in possession of each. The cost at which each is attainable is considered. Finally, a conception of the acquisition and possession of one of the objects yields intenser pleasure in the Emotive System, than is wrought by that of the acquisition and possession of any of the others. In its turn Emotivity operates simultaneously on Intellect and Acquisitivity. Intellect judges decisively and, at the same instant, the corresponding impulse leaps to operativeness. Here we have a judgment, constituting the intellective element of choice. We have, also, an anticipation, in which Imagination grasps the completion of the chosen course of action, and sheds on Emotivity a foretaste of triumph. There is a victorious impulse. There is likewise a Consciousness of Power, to which both the victory of the impulse and the anticipation of success contribute. Moreover, when the operative impulse is made vehement by the prospect of obstacles to be overcome, the intuitive Sense of Personalty is excited to extreme vividness. All these elements enter into the mental state; and who is prepared to say that such a blending cannot produce the feeling of which we are conscious at the moment of volition, or when a purpose is engaging our thoughts?

Perhaps I ought to give one or two illustrations of the method adopted to prove that a desire cannot cause a volition. Locke, as approvingly quoted by Stewart, says: "A man whom I cannot deny may oblige me to use persuasions to another, which, at the same time I am speaking, I may wish not to prevail on him. In this case it is plain that the will and desire run counter. I will the action that tends one way, whilst my desire tends another, and that the direct contrary." Here it is assumed that a person can have only one desire at a time. But every man's hourly experience of conflicting desires proves the assumption false. It is true that, in the case above instanced, there is one desire which opposes the voluntary action; but is it not clear that there is another—the desire to accede to the wish of the one who requires the action—and that this desire is triumphant in the volition which precedes the use of persuasion? In every case, where there is manifestly a motive which tends to prevent what is done, there is quite as manifestly a stronger motive which tends to promote it. Prof. Haven says: "We often wish or desire what we do not will. The object of our desires may not be within the sphere of our volitions, may not be possible of attainment, may not depend, in any sense, on our wills. Or it may be something which reason and the law of right forbid, yet, nevertheless, an object of natural desire." All this is very true, but it proves only that there are some desires which do not bear fruit in volition, not, by any means, that a volition ever takes place without an impulse in harmony with it, or that it presupposes the existence of a separate faculty. All such allegations as I have quoted are perfectly consistent with my definition of a volition.

As a matter of literary convenience, it is well to have such a substantive as "Will," and to understand by it the Mind's capacity to experience volitions. For obvious reasons there is more need of such a name at this point of vital activity, than there is of one to designate the Mind's capacity to experience a transition of vital force from the Emotive System to the Acquisitive System. We should always bear in mind, however, that when we are speaking of "The Will" as determining intellective or corporeal exertion, we are speaking figuratively, and picturing it as controlling the forces of which it is only a passive medium.

My reading in this line has not been extensive, but I understand that, among theologians of a certain school, many eminent and able men maintain that no explanation of volitions is possible, beyond the ascription of them to a central Will— "self-moved, self-directed." If any man can fix his attention on a human faculty, and really believe it capable of moving without an impelling force, he has a power of which I am glad to know myself destitute. I have no desire to divest myself for one moment of that Intuition of Causation, without an assumption of whose veracity it is impossible to write a single sentence containing a transitive verb. Aside from this consideration, in view of the unquestionable facts, which stand

out in blazing light all around us, it seems to me absolutely
wonderful that any thinker should hold to the doctrine that
volitions are independent of desire and affection, instead of
being their consequences. Everybody constantly and inevitably assumes the contrary. All confidence in men and all
distrust of them are based upon the assumption that their
voluntary actions result from their habitual motives. If we
know what a man's ruling passion is, we predict unhesitat-
ingly what his conduct will be when an opportunity for the
gratification of that passion shall be presented. Again, we invariably assume that the impellents are causes of volition whenever we attempt to influence a person to act, or to refrain from
acting. No one ever dreamed of a direct address to the will.
We address the intellect, hoping through this to awaken certain emotions, and thus to produce in the impulsive systems
such effects as will secure the volitions we desire. We often
ask ourselves, " Why did I do that?" and, if the act in question was a recent one, we have no difficulty in pointing out
the motive or motives from which it sprung. When a man
endeavors to explain or palliate his conduct, he tries to show
that his volitions were caused by impellents less unworthy
than those to which they might be attributed.

If an appeal to Consciousness is made, I profess my readi-
ness to appear before that tribunal at any moment. I shall
insist, however, upon a true statement of the deliverances of
Consciousness; for nothing is more common than a misinter-
pretation of them.when an attempt is made to clothe them in
language. If Consciousness is said to testify that our volitions are uncaused, the only proper attitude is that of President
McCosh: "A .direct contradiction." I cannot say with that

very able thinker, to whom the present generation is, and future generations will be, so largely indebted, that "This is a subject on which Consciousness, considered in itself, says nothing, and can say nothing." On the contrary I am sure that I am often *conscious of the causation of volitions.* I believe that this is always the case when one of my impulses rises to pre-dominance over opposing impulses. I am then conscious of the exercise of controlling power by that impulse; and that exercise is neither more nor less than causation.

If I am asked, "Does not Consciousness testify that we are free agents?" I answer Yes; and I accept that testimony without a shadow of reservation. But let us try to get a dis-tinct view of this Consciousness of Freedom. What is it? It is simply one of the forms of Consciousness of Power. It takes its peculiar cast from a conviction of the absence of restraint. My consciousness tells me that no earthly power can hinder me from resolving on any line of action on which I may choose to resolve. But no man's consciousness ever testified that he could come to a resolution in opposition to his own strongest impulses. When I say, "I can will to go, or will to stay, just as I choose," I tell the exact truth; and I may, or may not, forget that there are impellents within me now, determining my *choice* as surely as any effect ever fol-lowed its cause. The freedom of which we are conscious is the freedom of our volitional power from external restraint— the freedom of the human soul to will and to do according to its own prevailing impulses. But if we are conscious of this freedom from external restraint, are we not sometimes equally conscious of the *internal* enslavement of our volitional power? There are myriads of men who are saying today, "I cannot

5

make up my mind to take that step; I wish I could; I know I ought to do so; I know it would be best for me, but somehow I cannot bring myself up to such a decision." The pain with which this confession is made is due to Consciousness of Weakness in volitional power and reveals the dependence of that power on the forces which lie back of it. Here, as everywhere else, the affirmations of Consciousness, truly interpreted, are in perfect accord with conclusions from accurate observation and just induction.

I suppose the idea of a "Will self-moved, self-directed" has been caught up, on account of very unnecessary fidgeting about the fact of responsibility. But what is gained by such an expedient? Will anyone maintain that the Will is self-created, or that we had a hand, before we were born, in determining what the habits of our wills should be? The fact itself of responsibility is an indubitable and a stupendous reality, of which we see evidences everywhere, within us and without us. I find only silliness in any attempt, by raising metaphysical difficulties, to dodge that reality or shut it out of sight, when we all know that there is no possibility of evading the punitive consequences of anything that is evil in our mental State or external conduct. But I attach responsibility to the entire mental nature, and to every part of it. Though evil at one point is far worse, both in itself and in its consequences, than evil at another point, yet the law, *that badness shall yield suffering*, bears upon every power and upon every process belonging to mental life.

I am not in sympathy with those who seem nervously anxious to clear up the credit of the Supreme Mind; for I know with a perfect knowledge that the Order of the Universe is as it should be. While there are a thousand things in it which I shall not understand very soon, I remember that darkness has its uses as well as light. It is true that when we look into the human Mind, so beautifully organized for the evolution of Goodness and for the origination and diffusion of Happiness, we find it in sad disorder,—

" Like sweet bells jangled, out of tune and harsh."

It has seemed good that we should begin our careers at a stage where mental disharmony is causing manifold and multitudinous signs of badness and distress to emerge; and I can say, in all sincerity, that I am glad my initial lot has been cast just here. It is probable that, if we ever become able to measure the value of an enduring possibility of curing ills, rectifying wrongs and struggling upward, the existence of Evil will not then appear very mysterious.

CHAPTER VI.

A SENTIMENT, as I use the word, is a phenomenon to which both the Intellective System and the Heart contribute elements. It always contains a conviction and an emotion, and sometimes an impulse. I give the same meaning to the word "Sense" when I use it in naming a Moral Sentiment. These definitions make it evident that I do not regard any of the Moral Sentiments as underived or simple. But this fact does not derogate from my estimation of their originality and supreme importance. On the contrary, it may serve to elevate our conception of their dignity. We must remember that a Mind, at its first self-manifestation in the human child, has not fully emerged from the embryonic state, and that each stage necessarily reached in normal development marks a nearer approach to completeness than is indicated by any of the preceding stages. Consequently, the highest attributes are the latest in emerging. Indeed, at whatever point in healthful and continuous mental growth we look, we find the latest manifestation of mental life indicating more clearly than any that have gone before it the measureless distance between the level of man and that of the lower animals. I now proceed to the examination of the most important Moral Sentiments.

1. *The Critical Sentiment.*

In contemplating the actions and dispositions of our

fellow-men we decide that they are good or bad, and experience, accordingly, either pleasure or pain. The emotional
part of this sentiment obviously springs from Susceptibility
to Perception of Goodness and Perception of Badness. The
intellective conviction, however, demands a brief discussion.

I am in accord with those who ascribe all convictions,
whatever their subject matter, to the Intellective System. I
find no such elementary power as " Moral Reason," but believe
the use of such an expression to have a very misleading tendency. It produces the conception of a separate and unerring
faculty for the discovery of moral qualities. While it is true
that the activity of Intellect is a necessary condition of each
Moral Sentiment, that Power operates on matters of morality
under the same laws which govern its action when it is
engaged on other subjects; it acquires no additional efficiency
when it enters the moral realm ; nor is it any nearer to infal-
libility in one department of activity than in another, when
neither the affirmations of Consciousness nor those of the
Intuitive Faculty are involved. It would seem to me quite as
proper to speak of a Political Reason, or a Scientific Reason,
as to recognize the existence of a Moral Reason. I have
thought, sometimes, however, that an exhaustive exploration
of the Intellective System would disclose a Faculty which
would be fitly called the Critical Faculty, always busying
itself with the Susceptibilities, interpreting their contents,
conducting their education, and receiving education from
them. Such a faculty would have a relation to the Emotive
System similar to that of the Perceptive Faculty to the
Sensory System, and would have within its province all matters of taste, as well as all moral distinctions.

It may be best to say a word on some cavilings concerning this subject, although they seem almost too puerile to
deserve notice. Men have dwelt on the opposite views which
have prevailed in different ages, and among different nations,
concerning the rightness and wrongness of particular acts and
customs, and have argued from these disagreements that our
moral nature is an empty figment. No attempt at reasoning
was ever more inconsequential. It would be quite as rational
to infer, from the opposing scientific theories which have been
held, that man has no capacity for science. As I have heretofore shown, in all judgments, respecting goodness or
badness, qualities are first discovered by Intellect; Emotivity
applies the standard of pain or pleasure, and, in accordance
with the issue of this trial, Intellect, with or without delibera-
tion, classes the qualities as good or bad. If more remains
to be discovered, and attention is still fixed on the same
object, new apprehensions cause new results in the Emotive
System, and the classification of the qualities is correspond-
ingly modified. The process is one and the same, whether
the discrimination concerns morality, or has not the slightest
relation to it. Diversities in Moral Judgment prove only
that the Critical Faculty of one person has been more highly
educated than that of another.

I call attention here to a law which holds good in the
origination of all Moral Sentiments. It is analogous to that
noted and benign principle of the English Common Law,
according to which "Every man is presumed to be innocent
until he is proved to be guilty"—a principle which it would
be well for us to observe in all our judgments concerning our
fellow-men. We are under a structural necessity to observe it

in our judgments concerning ourselves. We necessarily presume that our emotions, impulses, volitions and actions are good until we are brought to see that they are bad. Consequently, our convictions as to the goodness or badness of that which we observe are, in the first instance, always in harmony with the emotions awakened by the observation, although a severe discipline of the Susceptibilities is often necessary, in order that new apprehensions may have their due effect in the Emotive System. We ascribe goodness to that which gives us pleasure, and badness to that which gives us pain, until it is proved to us that our emotions are unjustifiable. It is convenient to have a specific name for this law, and I call it the Law of Favorable Presumption. When the convictions, produced under it, are necessary and indestructible, they have all the dignity and are entitled to all the authority of intuitions.

The growth of our power to discern between good and evil depends immensely, of course, on circumstances; but, in every normally constituted mind, it must go forward for a period at some pace, and result in the recognition of many forms of goodness. Some of these forms are self-authenticating, and we can give no account of our favorable judgment on them beyond saying that we class them as good by reason of structural necessity. This is largely true, I think, as to physical beauty. We are so constituted that certain Sensory impressions, of which Intellect ascribes the origination to certain natural objects, excite pleasure in our Emotive System, and the intellective classification of those objects as good is a necessary consequence. A similar classification of corporeal strength and agility is provided for in Susceptibility to Consciousness of Power; and chiefly to the same origin we ascribe the universal conviction of the goodness of quickness of perception, tenacity of memory, shrewdness in planning and efficiency in expressing thought. Passing to qualities of the Heart, we see at once that a favorable judgment upon Benevolence is inevitable to the human mind. That judgment is necessitated by the reaction upon Intellect of Susceptibility to Perception of Happiness, because, as soon as the relation of cause and effect can be apprehended, Benevolence becomes inseparably associated in the young Mind with the production of pleasure. In this, as in all other critical judgments, the conviction is greatly strengthened by the activity of imagination ; the perceiver places himself in the position of the one toward whom a benevolent disposition, or its opposite, is manifested and shares the pleasure or pain which it tends to produce. Thus the conviction of the goodness of benevolence, and of the badness of its opposite, becomes as profound and indestructible as any conviction of which the human Mind is capable. The excellence of Courage also, is sure to be perceived whenever an exhibition of it is contemplated. Imagination, for the moment, places the observing Mind in possession of that quality ; and Consciousness of Power bears fruit at once in a favorable judgment.

A general conviction of the excellence of Purity seems to presuppose a considerable advancement in civilization. One of the most wonderful of all historical events was the promulgation of a ceremonial law having for a chief distinctive feature the education of a race of men to an appreciation of mental purity; and there are those who believe that the civilized world is, today, largely indebted to that ceremonial law and the educational efficacy realized in it ages ago for its deep conviction of the matchless worth of this virtue. The requisite stage of advancement being pre-supposed, the idea of purity is probably transferred from matter to Mind. Then it addresses the Emotive System through Perception of Goodness, calling in the testimony of no other Susceptibility, but simply asserting its own superlative excellence. It is an indispensable element of that form of Goodness which we know as beauty; and the Mind which apprehends it must commend it.

"We needs must love the highest when we see it."

I think no one will deny that men universally recognize the Goodness of Rectitude, or the purpose to do right. It is proper, however, to attempt the clearing away of obscurity at this point. In consequence of the tendency of moralists to dwell on external acts rather than on springs of action, they have employed the terms "Right" and "Wrong" far more frequently than "Good" and "Bad." They use them, too, as adjectives, as adverbs and as substantives; and no little confusion has arisen from this practice. I think it would be better to denominate the qualities of which these words are descriptive, "Rightness" and "Wrongness." Still an ambignity would remain. At one time we say an act is right because it is suitable to a relation; at another, because it is in accordance with an obligation. Oftener still these two reasons co-exist in our minds, sometimes the one, and sometimes the other, being uppermost. The two senses easily glide into each other since what is morally suitable is likewise incumbent as a duty. But the chief error consists in the ascription to rightness of a disproportionate importance.

It has often been maintained and still oftener assumed that this is the only quality with which morality has to do. In opposition to this view I hold that, whether we use the word "right" to express the idea of fitness, or that of consonance with obligation, we simply describe one of the forms of goodness, as we do, also, when we speak of an act as beautiful, or noble, or magnanimous, or sublime. Indeed, I am prepared to go farther than this and to say that rightness, conceived as the performance of duty, is a lower form of excellence than that of which any one of those epithets is descriptive. It is easy, however, to explain why "The Right" has received such prominence in works on moral philosophy. The fact of obligation is the one which first confronts us when moral improvement is under consideration. In a disordered soul, loyalty to obligation must precede spontaneous goodness. But thinkers have erred in not going back of this to discover the origin of Sense of Obligation, or forward of it to seek after the end toward which that principle is designed to conduct us. As a consequence, they have regarded the phenomena connected with the idea of Duty as ultimate, and have treated the purpose to do right as the sum of all morality.

I shall trace the genesis of Sense of Obligation on future pages. .It is safe here to assume its universality ; and this being admitted a single step will take us to the conclusion that the excellence of Rectitude is universally recognized. The idea of goodness necessarily enters into that of the discharge of obligation, and the idea of badness into that of its repudiation. We believe it our duty to perform one act because it is good, and to refrain from another because it is bad ; and the quality of goodness or of badness is neces-sarily transferred from the act to the purpose to act, or to refrain from acting. Hence, when by " Right " we mean consonance with obligation, we are inevitably convinced of the goodness of the purpose to do right. When the term is descriptive of suitableness to a relation the conviction comes from another source, but no less certainly. Fitness is one of the self-authenticating forms of goodness. It stands in need of no auxiliary, but compels approbation by its own inherent force. In this case, therefore, the transference of the quality from act to purpose ensures the classing of Rectitude as good.

While I maintain that these and other forms of moral excellence come of necessity to be viewed approvingly, I am far from asserting that such must continue to be the case. The intellective element in these judgments is ineradic-able ; but it is possible for a Mind to sink so far below its original state that this very belief will become a source of dislike. The Consciousness of Badness and comparisons un-favorable to itself may cause it to hate what it once loved. Thus the knave hates honesty, the miser detests benevolence, and the prostitute is enraged at the thought of chastity, although it is impossible for them, by any ingenuity in self-delusion, or by the most frantic vehemence of volition, to cast from themselves the knowledge that the virtues which awaken such antagonism are worthy of commendation.

2. *Sense of Justice.*

In contemplating some actions as manifestations of per-sonal qualities, we regard them as not only good or bad, but, also, as establishing a title to reward or a liability to just punishment, and are conscious of an impulse to favor the bestowment of the reward or the infliction of the punishment which we think to be merited. The obvious sources of this sentiment are Susceptibility to Perception of Goodness and Perception of Badness, Reactivity, and the Law of Favorable Presumption. We are pleased or pained by that which we observe ; we have an impulse to reciprocate the pleasure or pain experienced, and we presume that this impulse is justifiable. The favorable impulse is Complacency and the unfavorable one is Indignation.

It is plain that the growth of this sentiment must keep pace with the growth of the Critical Faculty. At first everything which causes a considerable degree of pleasure or pain gives rise to a disposition to reward or punish. The infant strikes the stick by which it is hurt, and fondles the toy by which it is pleased. Soon, however, it learns that only living beings are objects of just reward and punishment. Afterward it discriminates between intentional and accidental benefit and harm. It learns, also, to make allowance for lack of power and lack of knowledge. Then it discovers that pain may be inflicted for a good purpose and pleasure for a bad one. Finally, it learns that an act or a disposition may be of good or ill desert while it has no immediate connection with benevolence or its opposite. Thus a point is soon reached at which there are no farther distinctions of this nature to be made. Still, these distinctions are often overlooked even by adults. Indeed, there are few, if any of us, who do not occasionally lose sight of them for a moment. Some of the exhibitions of a misdirected Sense of Justice are very absurd. A mother, for example, snatches up and shakes her little boy for falling and hurting himself.

Her parental affection causes her to feel acute pain at witnessing the boy's suffering; and, viewing him, for the instant, as the originator of that pain, she experiences a sudden impulse to punish him. If we watch ourselves closely we shall all perceive that we are liable to have our indignation excited, by their unhappiness, against those for whom our affection is deepest. All such facts should serve to impress us with the importance of keeping the Critical Faculty in constant training. In proportion as it is uneducated, or misdirected, Sense of Justice degenerates into mere revenge, or becomes obtuse, and we are prone to make consonance with our own desires the measure of goodness.

This sentiment, however it may be perverted, is plainly universal and indestructible. The impulse embraced in it is an inevitable effect of an Elementary Power in conjunction with an Original Susceptibility; and the conviction, which it involves, will endure until the necessary presumption in favor of the impulse shall be destroyed by the force of evidence.

At this point we see one of the most important offices of Resentment. There are many writings upon that subject; and I should single out from among them those of Bishop Butler as having more value for me than all others that I have read. His allusion to human Resentment as an instrument of Divine Justice opens to our view a throng of realities which have an awful significance. He failed to observe however, that without this affection there could have been no such virtue as *human* Justice; that, consequently, human character would have lacked its most commanding feature, and that human society would have been destitute of its mightiest safeguard.

I will permit myself here to turn aside a moment, for the purpose of noticing that search after a "Final Cause" which is indicated in many discussions of the affections. What I am compelled to regard as a misguided Love of Unity has led philosophers to assume that the distinctive design of each affection is limited to a single end. When, therefore, they have discovered one of the offices of an affection, they have considered their work completed. But, in reality, it is characteristic of Wisdom to provide for many ends through a single agency. All of these we cannot, without great presumption, assume that we are able to discover. There may be instances in which we can discern but one, though I am satisfied that we should cease to speak of "*The* Final Cause" of any constituent feature of the human Mind.

3. *Self-judgment.*

We now turn our attention upon the sentiment associated with reflections upon our own characters as illustrated by our external actions and by the features of our mental life. According as our view of these is favorable or unfavorable, we pass a commendatory or condemnatory judgment on ourselves. The former judgment is delightful; the latter painful; and both the pleasure and the pain are susceptible of countless modifications dependent on the peculiar coloring of our favorable or unfavorable view. We look on ourselves as honest or dishonest, as courageous or cowardly, as loyal or disloyal, as veracious or false, as benevolent or selfish, as magnanimous or mean, as noble or contemptible; and with every variation of the judgment we experience a corresponding change in the resultant emotion. But all such judgments, including both the intellective and the emotive elements, may be classed either under the head of Self-approbation, or under that of Remorse.

It is a peculiarity of this Sentiment that the conviction, which enters into it, precedes and causes the emotion, while the immediate *legitimate* effect of the pleasure or pain belonging to it is the stimulation of Desire of Goodness. The primary source of Self-judgment is Susceptibility to Consciousness of Goodness and Consciousness of Badness; and the strengthening of it is provided for in several different processes. The emotional part of it is intensified by the reflex action of Desire of Goodness, giving the result of an impulse gratified or an impulse thwarted. To the same end operate two Moral Sentiments of which I shall soon give account. Anticipation of Retribution comes in to gladden with its promises or to appal with its threatenings. Here, too, is felt the exhilarating or the terrible force of Sense of Obligation; and we congratulate ourselves on a duty fulfilled, or reproach ourselves for a duty violated. Finally, Sense of Justice presents us to ourselves as entitled to be rewarded or deserving to be punished. When we are disposed to grumble about the inherited disorder which prevails among our desires and affections, we should do well to think of all these amazing structural provisions for our impulsion toward the realization of internal peace.

The deliverances of the Critical Faculty are more liable to be erroneous in connection with Self-judgment than at any other point. Though the principal danger lies in the other direction, it is unquestionable that erroneous Remorse is sometimes experienced. A love, for example, which has become an avenue of pain in consequence of the suffering or death of its object, often re-acts in self-accusations. Thus a mother, whose little child has been taken from her, may be heard confessing guilt with which she is not chargeable because of the withholding of some attention which she had not seen to be needful, or because of some punishment which she had conscientiously administered. Again, a desponding tendency, sometimes occasioned by disease, and sometimes resulting from a constitutional lack of vitality, may cause one to take unduly dark views of one's own character. But, on the other hand, it is evident that a large majority of mankind think too favorably of themselves. Every hour of observation furnishes illustrations of the truth that the wish is often "father to the thought." We are always prone to believe what we desire to have true; and so unwelcome is the conviction of our own badness that we are inclined to ward it off with great persistency. We are apt to become dishonest with ourselves and skillful in distorting all evidence pertaining to our moral state. Indications of badness are overlooked, while those of goodness are immensely magnified. Moreover, we often try ourselves by a false standard. Instead of adopting for this purpose the highest conception of human character which all our means of information enable us to form, we measure ourselves by ourselves and compare ourselves among ourselves. If we find that we are not morally below the average of the community in which we live, or of the circle in which we move, we are apt to conclude that we are not very different from what we should be. If to these causes, all growing out of a vicious state of our impulses, we add the familiar principle that no force is accurately measured until it is resisted, and remember the Law of Favorable Presumption, we shall see very clearly that the cost of self-knowledge is in direct proportion to its superlative value. It will be plain, too, that anything approaching accuracy of self-knowledge presupposes the incessant influence of Desire of Goodness on the Critical Faculty.

4. *Anticipation of Retribution.*

When, in connection with our future welfare, we are contemplating an action or a line of conduct, which is either unmistakably good or unmistakably bad, we anticipate either enjoyment or suffering as a consequence of that action or that line of conduct. This immediate effect of the contemplation is realized whether the action or conduct lies in the past, or is conceived as belonging to the future. According to this description, the activity of this Sentiment presupposes some measure of practical wisdom. Like other Moral Sentiments, Anticipation of Retribution is rendered inoperative for the time being by that foolishness, begotten of selfishness, which causes all remote consequences to be shut from view. Still, the Sentiment is indestructible, because there is a limit to the possible paralyzation of Desire of Happiness; and when once that forecasting desire is re-awakened, the vehemence of its activity is commensurate with the oppression to which it has been subjected, and the consequent recoil from the prospect of suffering is commensurately full of torture. A man, by utter recklessness, may go so far in the way of self-destruction as to think but seldom of any consequences of his conduct beyond the satisfaction of the mad impulses that are driving him on. But, whatever speculative fogs he may have gathered about his understanding, and however angrily he may say to himself that there is no such thing as Retribu-

6

tion, the time will never come when he can deliberately resolve on an evil course, with a full view of the evil that is in
it, and not know that he is resolving foolishly and acting the
part of an enemy to himself.

The origination of Anticipation of Retribution is necessitated by the very nature of our susceptibilities and desires.
A susceptibility is simply a power to experience pleasure and
pain; and all the knowledge, possessed in the first instance,
of the Primary Media of Pleasure, Knowing, Power, Ownership, Sympathy, Favorable Regard, Goodness, Foreign Goodness and Foreign Happiness—is derived from the pleasure
they yield; while the Primary Media of Pain—Not-Knowing, Weakness, Destitution, Lack of Sympathy, Unfavorable
Regard, Badness, Foreign Badness and Foreign Unhappiness
—are known only by the pain which they cause. Of necessity, therefore, the Media of Pleasure are conceived as sources
of enjoyment, and the Media of Pain take their places in the
Intellective System as sources of suffering. The two classes
of Media retain their respective characters when the idea of
"what is best on the whole" is evolved and Desire of Happiness comes into being; and they stand thenceforth, respectively, as sources of happiness and sources of unhappiness.
By the spontaneous action of Acquisitivity, the Primary
Media of Pleasure, as sources of happiness, become the respective objects of the Primary Desires, and these desires are
stimulated by the recoil of the Heart from the Primary
Media of Pain as sources of unhappiness. Thus it is made
impossible for the Mind ever to shake off the true conviction,
that the several Primary Media of Pleasure are promotive of
happiness and that their respective opposites are promotive

of unhappiness. It is needless to say that what is true of all these Media, of both classes, must be true of each one of them and, therefore, true of Goodness and Badness. Much more than this : we can all testify that, although the presence of other Media of Pain may cause us a certain feeling of humiliation, it excites no emotion of that nature comparable as to intensity with that which enters into the Remorse we experience from Consciousness of Moral Badness. Consequently, we know that the structural provisions for the conviction which connects Badness with the causation of unhappiness, are far more potent than those which yield the conviction of a like causative force in other Media of Pain. These considerations make it unquestionable that Anticipation of Retribution is a necessary, as well as an indestructible, Sentiment. Hence, it is entitled to all the evidential weight of an intuition. The Sentiment, in most cases, is strengthened immensely by the conception of a Just Moral Government, a conception for the evolution of which all the Moral Sentiments co-operate. Very often, indeed, it takes its color chiefly from that conception, while the idea of causation subsides.

A justification of Anticipation of Retribution, by the inductive method, is an easy matter for any one who will be at the pains to observe what is taking place within him and what is going on around him.

5. *Sense of Obligation.*

In pre-considering a possible action or possible series of actions, whether it be external or purely mental, we are convinced of an obligation to act or to refrain from acting, and have an impulse, more or less strong, toward such acting or refraining. This sentiment necessarily involves the recognition of authority, and is characterized by imperativeness. In expressing it we use the words "Right" and "Wrong," "Obligation," "Duty," "Ought," and sometimes "Must." We have the same conviction in connection with the possible actions of others, and often, though not necessarily, we desire the fulfillment of the apprehended or supposed obligation.

To account for this Sentiment, we must scrutinize the fact of obligation ; and we find a clue to its contents in the circumstance that we sometimes express it by the word "Must." This word is suggestive of necessity ; and, paradoxical as it may appear, it is nevertheless true that we often recognize the necessity of a free choice. It is not uncommon for us to say of a voluntary action, "It was very unpleasant for me, but I *had* to do it ; I couldn't help it." This recognition is not always associated with Sense of Obligation, but may, on the contrary, be diametrically opposed to it. One may be heard saying, "I know that I ought not to take that course, but, somehow, I *must*." It is easy enough to understand the state of mind thus indicated. Through self-indulgence and the law of Growth by Exercise, some desire, entirely commendable within its legitimate sphere, has become so inordinately strong as to predominate over Desire of Goodness. This may be called a Selfish recognition of Necessity. But there is on the other hand a Self-denying recognition of Necessity, in which we are moved to say, "It is my duty to do it, and I *must* do it whatever the consequences." Now, in such a case as this two circumstances are always present. In the first place, Desire of Goodness has become operative, and, in the second place, that operativeness has been achieved through a conflict with powerful opposing impulses. Such cases are frequent in the experience of every man who has not completely turned his back on personal goodness as an object of endeavor ; and it is plain that in this self-denying recognition of necessity, Desire of Goodness becomes imperative, and asserts its right to exercise dominion over all other impulses. Or, to speak without a figure, there is an actual exercise of dominion by Desire of Goodness and under the law of Favorable Presumption, there arises a conviction of the rightfulness of that dominion. It is true that, by reason of the same law there may be a temporary conviction of the rightful authority of a selfish impulse. Indeed, it is possible for a man to submit so abjectly and for so long a period to enslavement by selfish desire, that his debauched understanding will regard obedience to the enslaving impulse in the light of a duty. Let us suppose that he is both avaricious and strongly inclined to expensive self-indulgence. So long as the two forms of selfishness nearly balance each other, he says " I *ought* not to spend my money in this way;" and he uses this language because he sees that the objects of avarice are more enduring than those of the inclination to self-indulgence, and, therefore, that the former impulse is better entitled than the latter to furnish a rule of conduct. At length, by vigorous self-discipline, he raises avarice to the position of a ruling passion, and then, whenever the opposing inclination becomes violent, he says, " I *must* not fool away my money." Indeed, he often uses this language when resisting the highest impulses of which he is capable. He has consented to the supremacy of avarice and permitted it to prescribe a law to which he habitually ascribes a rightful authoritativeness. It is in the course of conflicts between impulses that his avarice has become imperative, and in consequence of such conflicts that he employs the words properly expressive of Sense of Obligation, and resolves on obedience to a law.

The process just traced may serve to illustrate the manner in which the Law of Goodness comes to be recognized. There is, however, an immeasurable difference between the two cases. The presumption in favor of the supremacy of the selfish desire is destroyed in the first moment of honest scrutiny; the conviction of rightfulness, therefore, falls away, and the law is seen to have no foundation. On the other hand, the more we investigate the grounds of our conviction, that the supremacy of Desire of Goodness is rightful, the stronger and more vivid it becomes, and the more commanding in our view are the sanctions with which the Law of Goodness is clothed. In the one case the conviction is fortuitous and easily destroyed; in the other it is necessary and ineradicable. There is a vast difference, also, between the penalties conceived in connection, respectively, with violations of the two laws. The avaricious man fears only the loss of money as a consequence of disobedience to the law which he has made to himself, while the man whose other impulses are dominated by Desire of Goodness, fears only the penalty of personal badness. This difference is seen most clearly when the two men are contemplating one and the same action. We may hear the one say, "I must pay this debt, or I shall be sued and subjected to costs," while the other says, "I must pay this debt, or I shall be dishonest." To be sure, in the latter case, Anticipation of Retribution is apt to come in with a menace of future suffering; but personal badness is the only penalty of which a conception may properly be said to enter into Sense of Obligation. Moreover, the avaricious man knows that the law which he obeys is one which he has created for himself and for himself alone. He may have so mutilated his understanding as to be, much of the time, under an illusory impression of obedience to a law imposed from above. But, whenever he looks directly at the nature of his law, he cannot fail to see that it is simply a rule of conduct which he has adopted only for himself. On the other hand, we cannot conceive the Law of Goodness as having been created by ourselves, or as a law whose authoritativeness is limited to ourselves. We are conscious of recognizing a law already in existence ; and it is impossible for us to believe that any rational created being is exempt from the obligations of that law. Let me say here that by "recognizing" a law, I mean simply apprehending its existence. We are compelled to believe that the Law of Goodness is, and always will be, in force, whether we purpose to obey it or not. This compulsory recognition arises from that necessary conviction of the inseparableness of Goodness and Happiness, as cause and effect, of which I have traced the origin in accounting for Anticipation of Retribution, together with the imperativeness of which we become conscious through conflicts of Desire of Goodness with selfish impulses. We are forced to believe that Goodness *must* yield enjoyment, and that Badness *must* yield suffering ; and, by reason of the Intuition of Causation, we recognize a cause of that necessity. Very properly we give to that cause the name of "Law"; for the idea of something by virtue of which a necessity exists is fundamental in our conceptions of Law, whether our thoughts are in the physical, the mental, or the moral realm. Although, as we have just seen, ideas of enjoyment and suffering are immediately concerned in necessitating the recognition of the Law of Goodness, I still maintain that they do not necessarily enter into Sense of Obligation. The Moral Sentiments have their several functions, but have nothing to say concerning the factors that have been operative in their origination. After much reflection on the matter, I am very sure that, during the sole activity of Sense of Obligation, we see Goodness as an end in itself, and have no thought of ulterior benefit.

I think I have now traced all the contents of the Sentiment under consideration to their sources, and am prepared to give the following definition : Sense of Obligation is Desire of Goodness, made imperative by the opposition of selfish impulses, and associated with the recognition of the Law of Goodness as eternal and universally authoritative.

I have been, at times, very near concluding that this Sentiment involves the recognition of a personal Ruler, as well as of a Law. This view may, at the first glance, seem to be supported by some of the terms we use. It may appear difficult to think of Obligation, or Duty, apart from a personal Being entitled to demand fulfillment; and it is plain that the use of these terms involves an assumption of the existence of such a Being. I have no doubt, indeed, that "Obligation" and "Duty" first came to be used in consequence of the conception of One possessing a supreme right to require Goodness in character and conduct. Nevertheless, by experiment and observation, we find it unquestionable that the relation indicated by these words,— the owing of obedience to the Law of Goodness — can be, and, in fact, habitually is, so abstracted from all that is above it as to be a temporary terminus of thought. Hence, it is only in consequence of a premature suspension of analytic labor that the recognition of a Ruler is ever supposed to belong to the essence of Sense of Obligation.

Although it is not strictly pertinent to the aim of this chapter, I think it best to say that I am as confident of the existence of a structural necessity to recognize Deity, as I am of any fact pertaining to the human Mind. It is true that there are able men, whose vocation embraces a constant observation of the operations of law, who have much to say concerning the immutability of all laws, who, in every sentence they utter, as well as at every step in their investigations, are compelled to assume the veracity of Intuition of Causation, and who yet seem desirous to have us believe that they find no place in the universe for a Lawmaker, but regard law itself as the ultimate goal of human thinking, and a barrier above which the Intuition of Causation never lifts itself. It is obvious that these gentlemen are unaware of what has happened to them. Had they observed themselves as faithfully as they have studied the things that can be seen and handled, they would never have blundered so enormously. Of course, it is easy enough to "*forget* God." That, indeed, is our overshadowing danger. It is easy enough for us to construct barriers to thought, and to form the habit of arresting our thinking at such artificial obstructions. In fact, such a habit inevitably grows up whenever any matter is for a long time of absorbing interest, and there is no operative motive to pass beyond it. Thus to the miser money is the "Be-all and the end-all," beyond which his thoughts never travel. The explanation is found in the mastery of thoughts by impulses together with the law of Growth by Exercise. It is by virtue of these forces that all intellectual habits are formed; and it is precisely this that has been experienced by those to whom matter and its laws have become the universe. They have a right, so far as we are concerned, to felicitate themselves on the paralysis of their highest faculties ; but it is not wise to accept them as teachers concerning matters with which all their intellectual habits have contributed to unfit them for dealing. Let them teach as they may, however, the great body of cultivated men, as well as the masses of mankind, giving free course to the spontaneous activity of their faculties, will continue to view the Moral Law and all other laws and all forms of matter, in the light of *effects*, however potent they may see them to be, also, as secondary causes. And whether the chain of cause and effect, which they may contemplate, be long or short, they will find its beginning in the "Cause of causes." At that goal of Infinitude, and there alone, the restless Intuitive Faculty, if subject to no violence, can find repose. But it is not in this character alone that Deity is necessarily recognized. In connection with Anticipation of Retribution, it is not so much the causation of law, as its certain execution, that is present to one's thoughts. Both from that Sentiment and from the recognition of the Law of Goodness, commonly called the "Moral Law," there inevitably springs a conception of attributes which can be ascribed only to a personal Being. Justice, Voluntariness and an Intelligence from which nothing can be hidden are necessarily recognized both in the causation and in the administration of the Law. Thus is necessitated the conviction of the existence of a Supreme Ruler. Moreover, the Mind presents structural provisions for his recognition in three different ways, as an all-powerful Helper. When mighty forces rise before us threatening us with destruction, and we fairly see our impotence to withstand them, Desire of Happiness impels us to an invocation in which God is treated as the only Being who can deliver us. Again, Desire of Happiness, conjoined with Consciousness of Badness and Anticipation of Retribution moves us to cry for mercy, and for deliverance from that evil within ourselves which we see to be a cause of unimaginable suffering. Finally, amid the tortures of Remorse, connected with a just appreciation of the terrible force of evil dispositions, Desire of Goodness impels us to invoke the Almighty One for rescue from the doom of eternal Badness. I thoroughly believe that all adult human beings of sound mind have passed through all these experiences, and, therefore, that all such persons, by virtue of their own organization, have arrived necessarily at the recognition of God as the all-powerful, all-wise and voluntarily acting, First Cause, Ruler and Helper. "So that they are without excuse."

If these things are true, it follows that God is not rightly named "The Unknowable." I freely admit, however, that the Agnostic philosophers have proved most abundantly that the human race stands in need of an external and a special revelation concerning Deity. It is well to remember, also, that some of the same philosophers, by their masterly disclosures of the correspondence, running through all the works of nature, between needs and provisions for satisfying them, have furnished materials for a weighty argument in support of the antecedent probability of such a revelation.

It will be observed that I have not employed the word

"Conscience" in discussing the Moral Sentiments. My reason is, that I saw in it only a source of confusion, and found no place for it with any meaning that I have ever seen attached to it. If used at all in scientific nomenclature, it should be restricted to Sense of Obligation. But I think it clearly best to dispense with it altogether as a scientific term, and let it continue to stand in popular literature for a conventional embodiment of all the Moral Sentiments, together with the moral function of the Critical Faculty.

An exhaustive examination of the Heart of Man would extend to the sentiments, embraced under the head of "Natural Religion"; but the discussion of these is not comprehended in my present design.

CONCLUSION.

I have finished the task which I proposed to myself. Whether my views will, or will not, receive any consideration, is a matter concerning which I have kept myself entirely free from expectation. I believe myself to have made some discoveries, though, with my limited reading, I am unable to say to what extent I may have been anticipated in the communication of them. If there is important and newly ascertained truth set forth in these pages, it will get into circulation through some channel "in the fulness of time." I had these things to write, and I have written them; and I choose to print and distribute a small edition of my treatise. It is altogether certain that I shall be satisfied with the result. I have no doubt that I shall come to see some of the points, on which I have touched, in a new light, nor that, if I should re-write this paper a year hence, I should modify some of my statements. Still, I will be frank enough to say that I am in no doubtful frame of mind so far as my principal conclusions are concerned. For a good number of years, I have occasionally tested, at various points, what I venture to call my theory of the Mind, by a somewhat habitual observation of myself and the persons around me ; and the result has been uniformly satisfactory. As to the doctrine, that the human Mind is an organized Being with a life of surpassing complexity, and with structural provisions for an infinitely greater variety of changes, processes, interactions and blendings, than is provided for in any corporeal organism on Earth, any material alteration of my opinion strikes me as an impossible event. In my hours of most humiliating self-distrust, when I see the limitations of my faculties most clearly, and am most ashamed of my proneness to intellectual blundering, my convictions on that and some other fundamental points stand out with no mists of doubt around them. I am confident that the five Elementary Powers, and the five Systems in which they respectively inhere, are substantially as I have described them ; and I think I see, with a fair measure of distinctness, the various processes concerned in the origination of volitions, and in the geneses of the Moral Sentiments.

In pointing out the phenomena of mental life, I have had occasion to disclose but very few glimpses of the animal life co-ordinated with it ; and the conception of the former life as still going on, after the extinction of the latter, is infinitely easier to me now than it was when I first entered on these researches.

www.ingramcontent.com/pod-product-compliance
Lightning Source LLC
Chambersburg PA
CBHW021412090426
42742CB00009B/1113